This book is to be returned on or before
the last date stamped

Safe Use of Solvents

Organizing Committee
(IUPAC Commission on the Atmospheric Environment)

S. Luxon (Chairman)
A.J. Collings (Secretary)
A.P. Altschuller
M. Benarie
K. Birett
R.S. Brief
M. Fugaš
O.H. Killingmo
K. Leichnitz
R. Smith

INTERNATIONAL UNION OF PURE AND APPLIED CHEMISTRY

IUPAC Secretariat: Bank Court Chambers, 2-3 Pound Way,
Cowley Centre, Oxford OX4 3YF, UK

INTERNATIONAL UNION OF PURE AND APPLIED CHEMISTRY
(Applied Chemistry Division)

Safe Use of Solvents

Proceedings of the International Symposium on the Safe Use of Solvents held at the University of Sussex, Brighton, UK 23–27 March, 1982

Edited by

A. J. COLLINGS

The British Petroleum Co. Ltd, Group Occupational Health Centre, Sunbury-on-Thames, Middx, UK

S. G. LUXON

Safety and Health Consultants, 47 Weymouth St, London, UK

1982

ACADEMIC PRESS

A Subsidiary of Harcourt Brace Jovanovich, Publishers

London   New York
Paris   San Diego   San Francisco
São Paulo   Sydney   Tokyo   Toronto

ACADEMIC PRESS INC. (LONDON) LTD.
24/28 Oval Road,
London NW1

United States Edition published by
ACADEMIC PRESS INC.
111 Fifth Avenue
New York, New York 10003

Copyright © 1982 by
INTERNATIONAL UNION OF PURE AND APPLIED CHEMISTRY

IUPAC Secretariat: Bank Court Chambers, 2-3 Pound Way,
Cowley Centre, Oxford OX4 3YF, UK

All Rights Reserved
No part of this book may be reproduced in any form by photostat, microfilm, or
any other means, without written permission from the copyright holder

British Library Cataloguing in Publication Data
Safe use of solvents
 1. Solvents
 I. Collings, A. J.   II. Luxon, S. G.
 661'.807   QD544

ISBN 0-12-181250-2

LCCCN 82-72343

Printed in Great Britain by
St Edmundsbury Press, Bury St Edmunds, Suffolk

CONTRIBUTORS

J.M. ANSELL Materials Safety Department, Administrative and Research Center, GAF Corporation, Wayne, NJ 07470, USA
P. APOSTOLI Istituto di Medicina del Lavoro, Universita de Padova, Policlinico di Borgo Roma, 37134 Verona, Italy
A. BERLIN Health and Safety Directorate, EEC, Batiment 2 Monnet, Luxembourg-Kirchberg, Luxembourg
R.J. BOSMAN Essochem Europe Inc., Nieuve Nijverheidslaan 2, B-1920 Machelen, Belgium
F. BRUGNONE Istituto de Medicina del Lavoro, Universita de Padova, Policlinico di Borgo Roma, 37134 Verona, Italy
C. BRULE Council of Europe, BP 431 R6, F-67 006 Strasbourg Cedex, France
C.P. BURGESS Health and Safety Executive, 25 Chapel St, London NW1 2DT, UK
A. CERRATI Department of Toxicology, University of Milan, via Vanvitelli, 32 Milan, Italy
M.F. CLAYDON The British Petroleum Co. PLC, Group Occupational Health Centre, Chertsey Road, Sunbury-on-Thames, Middlesex TW16 7LN, UK
P.O. DROZ Institute of Occupational Medicine and Industrial Hygiene of the University of Lausanne, Lausanne, Switzerland
M.J. EVANS The British Petroleum Co. PLC, Group Occupational Health Centre, Chertsey Road, Sunbury-on-Thames, Middlesex TW16 7LN, UK
P.A. FRANCO Department of Toxicology, University of Milan, via Vanvitelli, 32 Milan, Italy
M. FUGAŠ Institute for Medical Research and Occupational Health, P.O. Box 291, 41001 Zagreb, Yugoslavia
J.W.A. de GRAAF AKZO Zout Chemie Nederland, P.O. Box 25, 7550 Hengelo(O), Netherlands
M. GRASSO Department of Toxicology, University of Milan, via Vanvitelli, 32 Milan, Italy

CONTRIBUTORS

P. GRASSO   The British Petroleum Co. PLC, Group Occupational Health Centre, Chertsey Road, Sunbury-on-Thames, Middlesex TW16 7LN, UK

E. GUBERAN   Occupational Health Service of the Canton of Geneva, Geneva, Switzerland

L. HULSEN   Institute for Hygiene and Epidemiology, J. Wytsmanstraat 14, 1050 Brussels, Belgium

G. JACOBS   Institute for Hygiene and Epidemiology, J. Wytsmanstraat 14, 1050 Brussels, Belgium

M.J. KEBLE   Department of Toxicology, University of Milan, via Vanvitelli, 32 Milan, Italy

M.M. KEY   University of Texas Health Science Center at Houston School of Public Health, P.O. Box 20186, Houston, Texas, USA

J.J. KONIETZKO   Institute for Industrial and Social Medicine, Johannes Gutenberg University, Obere Zahlbacher Str. 67, 6500 Mainz, FRG

U. KOSS   Drägerwerk AG, P.O. Box 1339, 2400 Lübeck, FRG

P. LEINSTER   The British Petroleum Co. PLC, Group Occupational Health Centre, Chertsey Road, Sunbury-on-Thames, Middlesex TW16 7LN, UK

S.G. LUXON   Safety and Health Consultants, 47 Weymouth Street, London W1, UK

M. MARTENS   Institute for Hygiene and Epidemiology, J. Wytsmanstraat 14, 1050 Brussels, Belgium

F.H. van MENSCH   AKZO Zout Chemie Nederland, P.O. Box 25, 7550 Hengelo(O), Netherlands

M.K.B. MOLYNEUX   Shell (UK) Ltd, Occupational Hygiene Unit, Carrington, Urmston, Manchester M31 4AJ, UK

A.M. MOSES   Division Medical Department, ICI plc (Mond Division), P.O. Box 13, The Heath, Runcorn, Cheshire WA7 4QF, UK

G. MOSSELMANS   Directorate for the Protection and Improvement of the Environment, Commission of the European Communities, Brussels, Belgium

C. NICOLE   Institute of Occupational Medicine and Industrial Hygiene of the University of Lausanne, Lausanne, Switzerland

L. PERBELLINI   Istituto di Medicina del Lavoro, Universita de Padova, Policlinico di Borgo Roma, 37134 Verona, Italy

C. PUNTRELLO   Department of Toxicology, University of Milan, via Vanvitelli, 32 Milan, Italy

I. RAGGI   Department of Toxicology, University of Milan, via Vanvitelli, 32 Milan, Italy

R.A. ROBINSON   Occupational Health Group, BP Chemicals Ltd, Belgrave House, 76 Buckingham Palace Road, London SW1W OSV, UK

CONTRIBUTORS

R.A. SCALA Exxon Corporation, Medicine and Environmental Health Department, Research and Environmental Health Division, P.O. Box 235, East Millstone, NJ 08873, USA

R.G. SMITH School of Public Health, Department of Environmental and Industrial Health, The University of Michigan, Ann Arbor, MI 48109, USA

D.A. STAFFORD Department of Microbiology, University College, Newport Road, Cardiff CF1 1XP, UK

C.N. THOMPSON Shell UK Ltd, Shell-Mex House, Strand, London WC2R ODX, UK

J.A. WALTON Draeger Safety, Sunnyside Road, Chesham, Bucks HP5 2AR, UK

B.P. WHIM Imperial Chemical Industries PLC, Mond Division, P.O. Box 19, Runcorn, Cheshire WA7 4LW, UK

S.A. YUHAS Jr Exxon Chemical Company, Solvents Technology Division, P.O. Box 536, Linden, NJ 07036, USA

ACKNOWLEDGEMENTS

The Organising Committee thank Shell International Chemical Co. Ltd., The British Petroleum Company PLC, and Academic Press for their support.

PREFACE

In modern industry solvents are used in a wide range of applications ranging from the extraction of petroleum products, foods, and natural products to use as coupling agents, in adhesives, surface coatings and as carriers for perfumes, essences, insecticides and consumer products.

Because of this wide usage government, industry and trade unions have an interest that solvents are manufactured, stored, transported, used and disposed of safely. One step has been an attempt to classify solvents so that their containers can be labelled. However, before a solvent can be labelled or used safely its physical, chemical and biological properties need to be assessed.

With many solvents, especially those with a low flash point, care has to be taken to ensure that flammable concentrations of vapours do not develop and that all sources of ignition are avoided. If a fire or an explosion does occur steps should have been taken to contain it and provide suitable fire fighting equipment.

Some solvents are chemically reactive and may partake in exothermic reactions or react on storage to give dangerous reaction products e.g. peroxides.

Less obvious are the biological properties a solvent may have. Although the results of simple defatting of the skin by solvents has been known for many years more recently certain solvents have been shown to affect the nervous system, liver, kidney or other internal organs under certain conditions. Some solvents can induce cancer in experimental animals after prolonged exposure.

Even when a low toxicity solvent is used worker exposure should be kept as low as practicable by engineering design, exhaust ventilation and work practice. In addition to these controls there is a need to monitor individual workers and the workplace to ensure that the controls are effective and that excessive exposure does not occur.

Criteria for the control of worker exposure to solvents have been set by various countries. Usually similar limits are set but occasional differences do occur.

After use solvents have to be disposed of safely and their effects on the general environment and waste disposal systems have to be considered.

This volume is the proceedings of the International Symposium on the Safe Use of Solvents. In it international specialists have reviewed the factors necessary for safe use. Papers are also included concerning problems with specific solvents, solvent abuse and the training of personnel controlling the safe use of solvents.

The editors thank the members of the programme committee for their help and guidance and the contributors for making their papers available in good time for publication.

June 1982   A.J. Collings
                                                     S.G. Luxon

CONTENTS

Contributors	v
Acknowledgements	viii
Preface	ix

I: OPENING SESSION

An Industry View on the Safe Use of Solvents
   C.N. Thompson   1

The Health and Safety at Work of Solvents at the European Community Level
   A. Berlin and G. Mosselmans   7

II: IDENTIFICATION AND CLASSIFICATION

Hydrocarbon Solvents in Food-Related End-Uses : Compliance with F.D.A. Requirements
   R.J. Bosman   25

A Computer Program for the Labelling of Dangerous Substances
   G. Jacobs, M. Martens and L. Hulsen   35

European Systems for the Labelling of Solvents
   S.G. Luxon   53

III: FIRE AND EXPLOSION

Safety in Handling/Storage of Solvents at Terminals
   S.A. Yuhas, Jr   61

Detection and Prevention of Explosive Hazards for Solvent Vapours
   U. Koss   67

IV: TOXICOLOGY OF SOLVENTS

Recent Developments in the Toxicology of Solvents
   R.A. *Scala*   75

Major Toxicological Features of Chlorinated
Solvents
   K.K. *Beutel*   105

Some Aspects of the Comparative Toxicology of
Chlorinated Solvents
   A.M. *Moses*   117

Carcinogenicity of Solvents
   P. *Grasso*   129

Protective Action of 4-Methyl Umbelliferone Against
the Intoxification of Carbon Tetrachloride
   A. *Cerrati, P.A. Franco, M. Grasso, M.J. Keble,
C. Puntrello and I. Raggi*   137

Abuse of Solvents in Europe
   C. *Brule*   141

Sniffing 1,1,1 - Trichloroethane: Simulation of
Two Fatal Cases
   P.O. *Droz, C. Nicole and E. Guberan*   153

V: RATIONALE OF SETTING OCCUPATIONAL STANDARDS FOR EXPOSURE TO SOLVENTS

The United States Approach to the Setting of Standards
   M.M. *Key*   161

The Federal German Approach to Setting Standards
   J.K. *Konietzko*   173

The United Kingdom Approach to the Setting of
Standards
   C.D. *Burgess*   181

VI: MONITORING OF EXPOSURE TO SOLVENTS

Use of Detector Tubes for Evaluating Mixed Solvent
Vapours
   J.A. *Walton*   193

The Use of Air Monitoring Badges for Health Protection
in Handling Chlorinated Solvents Mixtures
   F.H. *van Mensch and J.W.A. de Graaf*   209

Lung Uptake of Isopropanol in Industrial Workers
  F. Brugnone, L. Perbellini and P. Apostoli   219

VII: CONTROL OF SOLVENT EXPOSURE

The Control of Industrial Exposure to Solvents
  M.K.B. Molyneux   225

Halogenated Solvents in Industry: Control of Solvent Exposures
  B.P. Whim   239

Advice to Customers of the Safe Handling of Solvents
  R.A. Robinson   251

Factors Affecting the Use of Filtering Respirators For the Control of Short-Period Exposures to High Concentrations of Organic Vapours
  P. Leinster, M.J. Evans and M.F. Claydon   261

The Use of Solvent Substitution as a Method for Improving Health and Safety
  J.M. Ansell   269

VIII: ENVIRONMENTAL ASPECTS

Education and Training in the Controlled Use of Solvents
  R.G. Smith   283

Biological Treatment of Organic Compounds and Solvents
  D.A. Stafford   293

Environmental Monitoring of Solvent Exposure
  M. Fugaš   305

AN INDUSTRY VIEW ON THE SAFE USE OF SOLVENTS

C.N. Thompson, CBE, BSc, FInstPet, CChem, FRSC

Shell UK Ltd, Shell-Mex House, Strand, LONDON, WC2R 0DX

INTRODUCTION

As the former, in fact, the first Convenor of the IUPAC British National Company Associates Group, and a member of the British National Chemistry Committee, it is indeed a great honour and pleasure to accept your kind invitation to give this opening address for such a major International Symposium.

The importance of the subject is a bit daunting, given that, as your opening speaker, I am expected to give an industrial over-view on such a complex subject.

When I was a schoolboy, I remember being reprimanded for always starting my essays the same way. If the subject was "APPLES", so my opening sentence was "There are very many sorts of apples . . . "

It's the same with solvents. There are very many sorts of solvents, and they have been used since the earliest civilizations. When prehistoric man started to paint his caves, or to use pigments for "writing" he would have discovered that water was not always the most suitable carrying medium. Animal and vegetable fats and oils or natural exudations of hydrocarbons from the earth were undoubtedly used because they gave superior, and more permanent colouring for his animal pictures in cave-drawings.

As crafts and trades proliferated, so the search for suitable solvents would have resulted in new uses for these materials. Most likely ethanol from distillation of sugar ferments would have been used at an early stage. It was certainly well known all over Europe and elsewhere from very early times.

As I said before, the variety of solvents is immense. There are probably many substances that some of you might have difficulty in recognising as solvents. Apart from some exotic ones, like liquid xenon, what for example do liquid sulphur dioxide, furfural, and two-phase mixtures of propane and cresol have in common? (apart, that is, from oxygen atoms?) The answer is that they have all been used extensively in the petroleum industry for the solvent-refining of lubricating oil. The first (liquid SO_2) was used for many years in the Edeleanu process for solvent extraction of spindle, insulating and white oils, and the other two in the refining of engine lubricants.

The present widespread use of solvents in industry can be traced back to 1784, when Lavoisier first demonstrated the true composition of organic substances which led to the foundation of aromatic chemistry. This resulted eventually in the use of coal tar naphtha from London gas plants for dissolving rubber (1823).

The chief sources of aromatic and aliphatic hydrocarbon are coal and petroleum the latter being the largest provider.

USAGE

Solvents are essential to modern life because they provide an effective cost/performance basis for the processing, manufacture and formulation of many thousands of products and substances — paints, adhesives, pharmaceuticals, as well as industrial processes such as mineral extraction, degreasing etc.

Often there are no technical or economic alternatives e.g. vegetable oil extraction, inks, adhesives, paints.

To illustrate the diversity of use for solvents, let us have a look at the end-use breakdown of organic solvents in the EEC.

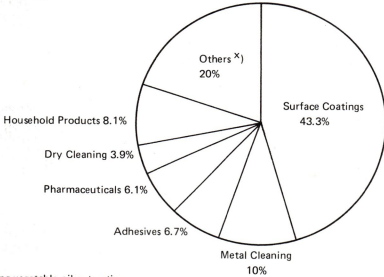

x) including vegetable oil extraction

Fig. 1. End-use breakdown of organic solvents — EEC

This is a typical breakdown for industrialised communities. It can be seen that by far the largest use for solvents is paints and other surface coatings.

CONSUMPTION

The consumption of solvents can be broken down further by type as we can see in Figure 2.

Industrial Solvents	Tonnes x 10^3	%
Aliphatic hydrocarbons	1200	28
Aromatic hydrocarbons	900	20
Chlorinated hydrocarbons	800	18
Alcohols	600	14
Esters	300	7
Ketones	450	10
Glycol ethers	150	3
	4400	100

Fig. 2. Consumption of industrial solvents in the E.E.C. (1980)

Almost half of the usage is accounted for by the plentiful and cheap aromatic and aliphatic solvents.

A growing use for ethanol is as a fuel ("gasohol"), or as a fuel component, particularly in parts of the world where it is readily available from natural product resources.

There are, of course, many non-solvent uses for solvents, the most prolific being the use of ethanol for human consumption (300 million litres per annum, world wide!).

HAZARDS

The main factors influencing solvent usage are:
a) Cost and availability
b) Performance and applicability
c) Environmental and health-related aspects.

Of this last, the hazard traditionally most often associated with solvents is their flammability. This is because the earliest used solvents were hydrocarbons which are notoriously easy to ignite, and burn very readily. The halogenated solvents, as well as the higher boiling hydrocarbons and oxygenated solvents, are generally less hazardous from this point of view.

The major hazard of fire arises when vapours from highly volatile solvents are allowed to build up to dangerous levels, either in processing operations, transport or storage. Where this is possible, smoking or naked lights must be rigorously excluded, and in such hazardous atmospheres, all electrical equipment, even the telephone, must be of the flameproof type designed for such circumstances. I am not thinking for example only of centrifuges and other plant used in the pharmaceuticals industry, but also of the use of things such as tools, and even torches, electronic watches, pocket calculators, some instant cameras and other motor-drive cameras that can also produce sparks capable of igniting unsuspected pockets of flammable solvent vapour. So too can static electricity, generated by splashing, and by over-fast flow-rates in pumping or transfer operations.

One recent handbook I have says:

"There are other ways in which ignition of a dangerous atmosphere can occur. Electrical storms can promote sparking, so no transfers (of flammable liquids) should be undertaken during thunderstorms, and dropping tools, covers, etc., on to concrete, rock or even brick, can cause sparks."

Other hazards have been well recognised for over 100 years. For example, nearly all solvents are narcotic in high concentrations and this property was used beneficially in the development of surgery when ether and chloroform were first used to produce anaesthesia (they were later abandoned because of toxic effects and other reasons). Dermatitis caused through de-fatting of the skin has also been recognised as a hazard involved in the use of certain solvents.

More recently other hazards have been recognised which are being well covered in other papers. Suffice it to mention just a few of the most serious:

Benzene — inhalation of benzene over long periods at low concentrations may cause:
a) Bone marrow damage leading to blood disorders of varying severity which in some cases is fortunately reversible after removal from exposure.
b) More rarely, aplastic anaemia or even possibly leukaemia which may occur long after exposure has ceased.

n—Hexane and Methyl n—butyl ketone — have been shown to exhibit neurotoxic effects — a peripheral neuropathy — after exposure to high vapour concentrations.

Glycol ethers — can adversely affect the blood, liver and kidneys and recent work is beginning to indicate that some of them may be able to cause reproductive defects.

Alcohols — the effects of abuse are so well known as to be hardly worthy of mention.

This recognition of health hazards has inevitably resulted in action by the producers, users and regulating authorities to limit exposure.

Most commonly this has been achieved through occupational exposure standards being set (Threshold Limit Values) which, if strictly observed, are a step along the road to eliminating these hazards.

This is a subject that will be well covered in Sessions V, VI and VII of the symposium.

In addition to health hazards the environmental effects of solvents in the atmosphere, in rivers and waterways and on land have to be considered. How one safely disposes of solvent wastes and seeks to avoid over-burdening the atmosphere with solvent vapours are all matters that have exercised the minds of industrial chemists and ecotoxicologists, as well as the authorities, for many years.

This too will be well covered in the symposium in section VIII.

INDUSTRY'S RESPONSIBILITIES AND ROLE

In the modern, developed world it would be a foolish and irresponsible industrialist who could consider it good and profitable business practice to harm his customers or the public at large. So at the outset industry has a vested interest in promoting health and safety both in the workplace and for the ultimate consumer. Moral and social considerations dictate that the supplier of a potentially dangerous substance should bear a responsibility to those who will come in contact with it.

But good intentions alone cannot remove from solvents their hazardous properties which are, in any case, intrinsic in their chemical structure and biological action. Exposure to any dangerous substance must be controlled. This presupposes that the hazards are known and recognised. This is where suppliers perform their biggest role in hazard-testing and research so that the nature and extent of the hazard can be established. The industrial user must then interpret the data to assess what significance it has for his own operations. An example would be the selection of a suitable solvent for vegetable oil extraction: it is not enough to know the toxicity of the solvent — the user has to establish how much of it, if any, will remain in the end-product, and decide whether this is likely to be harmful or not when the oil is consumed.

In protecting employees from harmful exposures to vapours, only the authorities can set the standards. This is not a job for industry but it must have a voice in indicating the nature of the problems and what is practicable. Unnecessarily restrictive standards can be as damaging to an employee's livelihood as ones that are too lax. Having got standards to work to, it is the function of management to implement them by providing safe systems of work. Designers, scientists and engineers also have their part to play in providing safe equipment (e.g. dry-cleaning machines that do not explode).

It is easy to list the hazards and the ways of combating them — be they the major hazards of large-scale production, storage and transportation or the hazards of solvents in use or to the consumer.

Every aspect of production, research and use calls for judgements to be made and this is what this conference will bring into focus. Needless to say health and safety are only achieved at a cost. The costs of the current and envisaged toxicological testing in the solvents industry, and the work to ensure that plant and equipment is safe to use, adds up to millions of pounds and every addition to our knowledge calls for additional costs, and further judgements.

So it is absolutely essential to maintain an appropriate sense of proportion, balancing the possible against the practicable, and the desirable against the reasonable. Having set the scene, as it is seen from an industrial standpoint, it is up to the conference itself, as a whole, in the papers and discussions that follow, to assist in furthering these worthy — indeed essential — aims.

In concluding, I would like to express my thanks to the many colleagues, inside and outside Shell, for their helpful discussions and provision of material. In particular, I should like to mention Peter Bourne and John Eberlein of Shell Chemicals UK Ltd, London, Leslie Turner of the Koninklijke/Shell Laboratories in Amsterdam, and Harry Lee of Glaxo Operations UK Ltd.

THE HEALTH AND SAFETY AT WORK ASPECTS OF SOLVENTS
AT THE EUROPEAN COMMUNITY LEVEL

A. Berlin and G. Mosselmans*

Health and Safety Director
Commission of the European Communities
Luxembourg

Directorate for the Protection and Improvement
of the Environment
*Commission of the European Communities
Brussels

1. INTRODUCTION

The protection of workers handling solvents has been approached through two different and almost independent pathways at Community level.
In the 1960's, the programme for the elimination of technical barriers to intra-community trade considered that there was a need for a unified approach towards the classification, packaging and labelling of dangerous substances. While the programme specifically excludes the use of such substances from its field of application, their classification with respect to danger for man, the specification of these hazards and the provision of safe-handling advice, furnish indirect but important elements for the protection of workers.
The 1978 action programme on Safety and Health at work (1), places special emphasis on the danger to workers exposed to chemicals; six of the fourteen actions of the programme call for harmonization of limit values, setting of hygiene measures, developing criteria and biological indicators. Among the toxic agents singled out for specific action are a number of solvents.

2. COMMUNITY LEGISLATION APPLICABLE TO SOLVENTS AT WORK

Two groups of related Directives elaborated within the framework of the Community aimed at eliminating technical barriers to trade provide some indirect degree of protection for workers exposed to solvents:

- The 1967 Directive on the classification, packaging and labelling of dangerous substances (2) and its sixth amendment (1979) (3).

- The 1973 Directive on the approximation of the laws, regulations and administrative provisions of the Member States relating to the classification, packaging and labelling of dangerous preparations (solvents) (4) and its 1980 amendment (5).

Within the framework of the action programme on Safety and Health at Work the Council Directive of 27 November 1980 on the protection of workers against the risks connected with an exposure to chemical, physical and biological agents at work (6) provides the first direct elements for active measures to be taken at Community level for the protection of workers against the harmful effects of certain solvents.

3. CLASSIFICATION OF DANGEROUS SUBSTANCES

As mentioned in section 2 the 1967 Directive is the first Community directive concerning dangerous substances. However, some historical dates should be recalled, which formed the origin of the 1967 Directive. In 1948 "Standard Safety regulations for government and other industrial establishments" prepared by the International Labour Organisation (ILO) were approved in Geneva. The provisions were amplified when on 21st April 1950 the Chemical Industries Committee of ILO adopted a resolution on classification, labelling and International safety symbols for dangerous, harmful and toxic chemical substances.

In 1955 the Social Committee of the Brussels Treaty Organisation (which does not exist any more) published a first list of toxic and dangerous chemicals with proposals concerning their labelling and using the symbols proposed by the Chemical Industries Committee:

Toxic:	a skull and cross bones
Harmful:	a St. Andrew's cross
Corrosive:	a symbol showing the damaging effect of an acid
Flammable:	a flame
Oxidizing:	a flame over a circle

EEC LEGISLATION ON SOLVENTS

In 1960 the Council of Europe carried out the work on this field and published in 1962 a list of 500 dangerous chemicals better known as "the Yellow Book" (1st edition: 1962 - 4th edition 1978).

The Recommendations concerning the labelling of those dangerous substances were addressed to the authorities of the countries which had ratified the "Partial Agreement in the social field".

In 1967 the Commission of the European Community submitted to the Council of Ministers a proposal, based on work done at international level, for a directive concerning the approximation of the laws, regulations and administrative provisions relating to the classification, packaging and labelling of dangerous substances.

The concept of dangerous substances has significantly evolved since 1967. With the adoption in 1979 of the Sixth Amendment (79/831/EEC), well known, amongst other measures, for its premarketing notification requirement of new chemicals, also contains provisions relating to classification and labelling to both new and existing substances (including imported chemicals) placed on the Community market.

The EEC labelling requirements are intended to provide a clear primary means by which all persons (workers as well as public at large) handling or using substances are given essential information about the inherent dangers of certain such materials. The means used area combination of symbols, standard risk phrases (R phrases) and standard safety advices (S phrases). The symbols highlight the most severe hazards presented by the substance; the R phrases try to give a more specific picture of those hazards, and the S phrases give safety advices on necessary precautions and/or of mishandling to be avoided relating to the use of the substance.

The EEC label is intended to give information on two types of dangers: health dangers and physical dangers. Fourteen definitions of dangers are given in the directive 79/831/EEC (article 2, para. 2). Of these fourteen types of dangers, nine are covered by symbols as described in article 16 para. 2c and annex II.

Definitions of dangers and corresponding symbols are listed in Table I.

TABLE 1

CLASSIFICATION AND LABELLING SYMBOLS
FOR DANGEROUS SUBSTANCES

CLASSIFICATION	LABELLING SYMBOL	
EXPLOSIVE	EXPLODING BOMB (E)	
OXIDIZING	FLAME OVER A CIRCLE (O)	
EXTREMELY FLAMMABLE	FLAME (F)	
HIGHLY FLAMMABLE	FLAME (F)	
VERY TOXIC	SKULL AND CROSS-BONES (T)	
TOXIC	SKULL AND CROSS-BONES (T)	

CLASSIFICATION	LABELLING SYMBOL	
HARMFUL	ST. ANDREW'S CROSS (XN)	
CORROSIVE	SYMBOL SHOWING THE DAMAGING EFFECT OF AN ACID (C)	
IRRITANT	ST. ANDREW'S CROSS (XI)	
DANGEROUS FOR THE ENVIRONMENT		
CARCINOGENIC		
TERATOGENIC		
MUTAGENIC		

At present there are 40 recognised single sentences and 17 combinations describing the nature of the special risks attaching to dangerous substances and 45 recognised single sentences and 10 combinations of 5 phrases giving safety advices.

Examples of these typical R and S phrases are given in Table II.

II a : Typical Risk phrases (examples related to toxicity)
II b : Typical Safety advice

TABLE II a

TYPICAL RISK PHRASES (R)

(Examples Related to Toxicity) (8)

R-21	HARMFUL IN CONTACT WITH SKIN
R-26	VERY TOXIC BY INHALATION
R-32	CONTACT WITH ACID LIBERATES VERY TOXIC GAS
R-33	DANGER OF CUMULATIVE EFFECTS
R-36	IRRITATING TO EYES
R-36/37	IRRITATING TO EYES AND RESPIRATORY SYSTEM

TABLE II b

TYPICAL SAFETY ADVICES (S PHRASES)

(Examples Related to Toxicity) (8)

S 4	KEEP AWAY FROM LIVING QUARTERS
S 5	KEEP CONTENTS UNDER...(APPROPRIATE LIQUID TO BE SPECIFIED BY MANUFACTURER)
S 21	WHEN USING DO NOT SMOKE
S 27	TAKE OFF IMMEDIATELY ALL CONTAMINATED CLOTHING
S 45	IN CASE OF ACCIDENT OR IF YOU FEEL UNWELL, SEEK MEDICAL ADVICE IMMEDIATELY (SHOW THE LABEL IF POSSIBLE)
S 36/37/39	WEAR SUITABLE PROTECTIVE CLOTHING, GLOVES AND EYE/FACE PROTECTION

The substances shall be classified as very toxic, toxic or harmful according to the following criteria (see Annex VI of Directive 79/831/EEC):

a) classification as very toxic, toxic or harmful shall be affected by determining the acute toxicity of the commercial substance in rat expressed in LD50 or LC50 value with the parameters as shown in Table III being taken as reference values.

b) if facts show:
  - that for the purpose of classification it is inadvisable to use the LD50 or LC50 values as principal basis because the substance produces other effects;
  - or the existence of effects other than the acute effects indicated by experiments with animals e.g. carcinogenic, mutagenic, allergenic, sub-acute or chronic effects substances shall be classified according to magnitude of their effects.

TABLE III

LD50 AND LC50 REQUIREMENTS FOR THE CLASSIFICATION OF TOXIC AND HARMFUL SUBSTANCES

Category	LD50 oral rat mg/kg	LD50 cutaneous rat or rabbit mg/kg	LC50 inhalation rat mg/liter/4hours
Very Toxic	$\leqslant 25$	$\leqslant 50$	$\leqslant 0.5$
Toxic		50-400	0.5-2
Harmful	200-2000	400-2000	2-20

Up to now close to one thousand substances have been examined, classified and listed in Annex 1 of the directive 67/548/EEC. This total of 1000 substances is however very small compared with the 50-60.000 existing commercial chemical substances amongst which at least 40% are dangerous in the meaning of the directive and for which sufficient physico chemical and/or toxicological data exist to permit their classification and labelling.

This lack of information is the main reason for the fundamental modification in approach which has taken place with the 6th amendment.
- For existing substances classification and labelling must take place in so far as the manufacturer or any person who place these substances on the market may reasonably

be expected to be aware of their dangerous properties. The data required for classification and labelling of these substances may have to be derived from different sources, for example previous test data, information required in relation to international rules of dangerous goods, information obtained from the literature or derived from practical experience.
- For <u>new substances</u> classification and labelling is mandatory and will be based on the data submitted to the Competent Authorities in the notification dossier. This dossier will include:
  - a technical dossier supplying the information necessary to evaluate the risks which the new substance may entail for man and the environment. It should contain at least the information and results of the studies referred in the so called base set (Annex VII) which concerns physicochemical, toxicological and ecotoxicological tests; Table V shows the toxicity tests as foreseen in the base set;
  - a declaration concerning the unfavourable effects of the substance in terms of the various uses envisaged;
  - the proposed classification and labelling;
  - the proposals for any recommended precautions relating to the use of the substance.

TABLE V

BASE SET OF TOXICITY TESTS

---

1. ACUTE TOXICITY

    1.1 LD50 ORAL, INHALATION, CUTANEOUS
        Usually two routes of administration
        Rats male and female
        14 days observation

    1.2 SKIN IRRITATION
        Albino rabbit

    1.3 EYE IRRITATION
        Rabbit

    1.4 SKIN SENSITIZATION
        Guinea-pig

2. SUB-ACUTE TOXICITY

    28 day administration. Usually oral.
    Rat preferable

3. MUTAGENICITY

    Series of two tests
    Bacteriological with and without metabolic
      activation
    non-bacteriological

It is clear that the results of these tests will allow not only a more appropriate classification of the substance if it presents dangers, and definition of these dangers, but will also serve as one of the sources of information for setting control requirements as well as hygiene and medical surveillance measures for workers exposed to these substances.

Up to now the toxicity and classification of substances was only considered when present in the pure form. Solvents are most of the time used as mixtures of variable composition. To cover this aspect of classification and labelling a Council Directive on preparations was first adopted in 1973 and subsequently modified on several occasions.

For example to determine if a preparation is to be classified and labelled as toxic or harmful, an empirical computation system has been set up based on a classification index for the toxic and harmful substances which may compose this preparation. The indexes to be used in this classification are summarised in Table VI, while the equations to be used for the computations are given in Table VII.

TABLE VI

CLASSIFICATION INDEXES FOR
TOXIC AND HARMFUL SUBSTANCES FOR THE CALCULATION
OF THE TOXICITY OF PREPARATIONS

Class of Substance		Classification Index I_1	Exemption Index I_2
Very Toxic and Toxic	I/a	500	500
	I/b	100	100
	I/c	25	25
Harmful	II/a	5	20
	b	2	8
	c	1	4
	d	0.5	2

TABLE VII

EQUATIONS FOR THE CALCULATION OF TOXICITY OF PREPARATIONS

1)		$\Sigma [P \times I_1] > 500$	Toxic
2)	and	$\Sigma [P \times I_1] \leq 500$ $\Sigma [P \times I_z] > 100$	Harmful
3)		$\Sigma [P \times I_z] < 100$	No Classification

P = % by weight of each substance
P_{min} class I 0.2%
       class II 1%

An indication of the number of substances classified in the various cagegories in 1973 and 1982 is given in Table VIII.

TABLE VIII

NUMBER OF TOXIC AND HARMFUL SUBSTANCES CLASSIFIED FOR THE PURPOSE OF COMPUTING THE TOXICITY OF PREPARATIONS

		1973	1982
Very Toxic and Toxic	I/a	7	11
	I/b	5	5
	I/c	6	4
Harmful	II/a	6	13
	II/b	13	12
	II/c	14	15
	II/d	12	16

All substances classified at present in category I are listed in Table IX.

Examples of the use of these indexes for the classification of preparations are given in Table X. In addition, of course, the manufacturer has to indicate if the preparation presents other dangers such as flammability or corrosivity and give safe-handling advice.

It must however be emphasized that these directives per se cannot formally contribute to the worker protection since they do not and cannot lead to any prohibition or limitation of use at work.

TABLE IX

SUBSTANCES CLASSIFIED IN CATEGORY 1 (1982) FOR THE CALCULATION OF TOXICITY OF PREPARATIONS

I/A	I/B	I/C
CARBON DISULPHIDE	BIS(2-CHLOROETHYL) ETHER	1-BROMOPROPANE
*BENZENE		METHANOL
CARBON TETRACHLORIDE	PHENOL	ACETONITRILE
1,1,2,2 TETRACHLOROETHANE	CRESOL	2-HEXANONE
PENTACHLOROETHANE	FURFURAL	
NITROBENZENE	* PIPERIDINE	
ANILINE		
*1,1,2,2 TETRABROMOETHANE		
*ALLYL ALCOHOL		
*1,2 DIBROMOETHANE		
*2-CHLOROETHANOL		

* Additions since 1973

A number of other substances have been downgraded: i.e. 1,1,2 trichloroethane (I/B → II/A), PYRIDINE (I/C → II/A).

4. PROTECTION OF WORKERS FROM CHEMICAL DANGERS AT WORK

The 1980 Council Directive (6) is a broad Framework Directive for the protection of workers from chemical dangers which should result in all Member States following a similar path in the future.

This Directive sets out two objectives:

- eliminate or limit exposure to chemical, physical and biological agents and prevent risks to workers' health and safety;

- protect workers who are likely to be exposed to these agents.

This Directive which will affect the majority of workers in the Community requires the Member States to take short term and longer term measures. The short term measures not applicable to solvents require that within three years workers and/or their representatives shall have access to appropriate information concerning asbestos, arsenic, cadmium lead and mercury, and within four years appropriate surveillance of the health of workers exposed to asbestos

TABLE X

EXAMPLES OF CLASSIFICATION OF SOLVENTS

		1,1,2,2 TETRACHLORO-ETHANE	1,1,2 TRI-CHLOROETHANE	1,2 DICHLO-ROETHANE	CYCLO-HEXANOL	$\Sigma(PI)$	CLASSIFICATION
I_1		500	5	2	0.5		
I_2		500	20	8	2		
% WEIGHT COMPOSI-TION (P)	1) 1%	1%	9%	10%	80%	$\Sigma PI_1 = 605$	TOXIC
	2) 0.5%	0.5%	0.5%	1.0%	98%	$\Sigma PI_1 = 300$ $\Sigma PI_2 = 464$	HARMFUL
	3) -	-	-	2%	98%	$\Sigma PI_1 = 51$ $\Sigma PI_2 = 212$	HARMFUL
	4) (0.1%)	(0.1%)	(0.5%)	9%	10%	$\Sigma PI_1 = 23$ $\Sigma PI_2 = 92$	NO CLASSIFICATION
	5) 0.3%	0.3%	(0.5%)	1.0%	10%	$\Sigma PI_1 = 157$ $\Sigma PI_2 = 178$	HARMFUL

and lead shall take place.

The longer term measures apply when a Member State adopts provisions concerning an agent. In order that the exposure of workers to agents is avoided or kept at as low level as is reasonably practicable, Member States must comply with a set of requirements and inform the Commission, but in doing so they have to determine whether and to what extent each of these requirements is applicable to the agent concerned. These requirements are the following:

- limitation of the use of the agent at the place of work
- limitation of the number of workers exposed or likely to be exposed
- prevention by engineering control
- establishment of limit values and of sampling procedures, measuring procedures and procedures for evaluating results
- protection measures involving the application of suitable working procedures and methods
- collective protection measures
- individual protection measures, where exposure cannot be avoided by the other means
- hygiene measures
- information for workers on the potential risks to which they are exposed, on the technical preventative measures to be observed by workers and on the precautions taken by the employer and to be taken by workers
- use of warning and safety signs
- surveillance of the workers' health
- keeping updated records of exposure levels, lists of workers exposed and medical records
- emergency measures for abnormal exposures
- if necessary, prohibition of part of all of the agent or agents involved, in cases when use of the other means available does not make it possible to ensure adequate protection

In addition an initial list of eleven agents, including 4 solvents (Table XI), relates to the implementation of further more specific requirements, which are as follows:

- providing medical surveillance of workers by a doctor prior to exposure and thereafter at regular intervals. In special cases, it shall be ensured that a suitable form of medical surveillance is available to workers who have been exposed to the agent after exposure has ceased;

- access by workers and/or their representatives at the workplace to the results of exposure measurements and, in the case of an agent for which such tests are laid down, to the anonymous collective results of the biological tests indicating exposures;

TABLE XI

SOLVENTS REQUIRING INDIVIDUAL DIRECTIVES

IN ACCORDANCE WITH THE 1980 COUNCIL DIRECTIVE (6)

ON THE PROTECTION OF WORKERS

SOLVENT	1973 SOLVENTS DIRECTIVE (AND AMENDMENT) CLASSIFICATION	IARC (1982) CLASSIFICATION
BENZENE	1/A	1
CARBON TETRACHLORIDE	1/A	2 B
CHLOROFORM	11/A	2 B
PARADICHLOROBENZENE	–	Insufficient Data for Evaluation

- access by each worker concerned to the results of his own biological tests indicating exposure;
- informing workers and/or their representatives at the workplace where the limit values are exceeded, of the causes thereof and of the measures taken or to be taken in order to rectify the situation;
- access by workers and/or their representatives at the workplace to appropriate information to improve their knowledge of the danger to which they are exposed.

This Directive also requires the Member States to consult the social partners when the above requirements are being established.

With regard to the eleven agents mentioned above the Council on proposal of the Commission will fix in individual Directives the limit value(s) as well as other rules.

On the basis of the experience gained with the first two proposals of Individual Directives, on lead and asbestos, now being discussed at Council an indication of the structure of these Directives which will most probably also apply to solvents is given in Table XII

TABLE XII
STRUCTURE OF INDIVIDUAL DIRECTIVES
FOR THE PROTECTION OF WORKERS EXPOSED TO
CHEMICAL AGENTS AT WORK

STRUCTURE	COMMENTS
1) AIMS OR OBJECTIVES	Political statement
2) DEFINITIONS	Terms, quantities, units
3) SCOPE	Coverage and exemptions
4) SPECIFIC CONDITIONS	Sometimes - prohibitions of certain practices
5) REPORTING PROVISIONS	Certain operations may require notification or authorization
6) GENERAL CONTROL PRINCIPLES	Main approaches to limit exposure
7) LIMITATION OF DOSES IN SPECIAL CASES	Critical groups of workers
8) ASSESSMENT OF EXPOSURE	Sampling and analytical strategy
9) EXPOSURE LIMITS	Numerical Values. Ambient if applicable biological
10) PLANNED SPECIAL EXPOSURES	For certain operations precautions required
11) EXCEEDING THE LIMIT VALUES	Provisions to be taken in such cases
12) INDIVIDUAL PROTECTION	Indicate conditions
13) RECORDS	Measurement of exposure health records
14) MEDICAL EXAMINATIONS	Frequency and guidelines
15) PERSONAL HYGIENE	Facilities and requirements
16) INFORMATION	Information of workers regarding dangers and precautions marking
17) STATISTICS	Information for monitoring application
18) APPLICATION	Date when operational

A special emphasis will be placed when possible, on the basis of scientific information and analytical practicability on the biological monitoring of workers exposed to solvents.

To assist with this task the Commission has started the publication of a series of monographs on the Human

Biological Monitoring of Industrial Chemicals. The first of these monographs devoted to benzene (7) recommends as biological indicators of exposure the monitoring of benzene in exhaled air, and in the case of higher levels of exposure and absence of exposure to phenols, the monitoring of phenol in urine.

5. CONCLUSIONS

We feel thus that within a few years the Community should have developed most instruments required by Member States to ensure that solvents are used in the safest manner possible by the tens of millions of workers in the European Community called daily to handle them. Special attention will still have to be paid to the teaching and academic research laboratories where often such solvents are used and where usually precautions are not so severe, and which are at present excluded from the field of application of these directives.

From the very early stages of teaching chemistry, toxicological concepts could be introduced so that the handling of chemicals and solvents in particular not be regarded casually as an innocuous matter.

BIBLIOGRAPHY

1) Council Resolution of 29 June 1978 on an action programme of the European Communities on Safety and Health at Work.
Official Journal of the European Communities C 165, pg 1, (11.7.1978).

2) 67/548/CEE Directive du Conseil du 27 Juin 1967 concernant le rapprochement des dispositions législatives, réglementaires et administratives relatives à la classification, l'emballage et l'étiquetage des substances dangereuses.
Journal Officiel des Communautés Européennes 196, pg 1, (16 Août 1967).

3) 79/831/EEC Council Directive of 18 September 1979 amending for the sixth time Directive 67/548/EEC on the approximation of the laws, regulations and administrative provisions relating to the classification, packaging and labelling of dangerous substances. Official Journal of the European Communities L 259, pg 10, (15.10.79).

4) 73/173/EEC Council Directive of 4 June 1973 on the approximation of Member States' laws, regulations and administrative provisions relating to the classification, packaging and labelling of dangerous preparations (solvents). Official Journal of the European Communities L 189, pg 7, (11.7.73).

5) 80/781/CEE Directive du Conseil du 22 Juillet 1980 modifiantla Directive 73/173CEE concernant le rapprochement des dispositions législatives, réglementaires et administratives des Etats Membres relatives à la classification, l'emballage et l'étiquetage des préparations dangereuses.
Journal Officiel des Communautes Europeennes L 229, pg 57, (30.8.1980).

6) 80/1107/EEC Council Directive of 27 November 1980 on the protection of workers from the risks related to exposure to chemical, physical and biological agents at work.
Official Journal of the European Communities L 327, pg 8, (3.12.1980).

7) Human Biological Monitoring of Industrial Chemicals: I Benzene, R. Lauwerys. EUR 6570 EN.

HYDROCARBON SOLVENTS IN FOOD-RELATED END-USES

COMPLIANCE WITH F.D.A. REQUIREMENTS

R.J. BOSMAN

*Essochem Europe Inc., Nieuwe Nijverheidslaan 2,
B-1920 Machelen, Belgium*

The industrial transformation of a crop into a food that can be eaten by man necessitates a large number of steps. In many of these steps, extensive use of chemical products is made. It is probably reasonable to assume that by the time it is ready for consumption, a food has seen, on the average, a dozen chemicals. Just to mention a few: fertilizers - pesticides - preservatives - cleaning agents - solvents.

In this chain of food-related uses, solvents form a class by themselves. I would like to review some of the features that make these solvents special. In all cases these features correspond to the need for these solvents to meet the stringent regulations established by the Governments. The purpose of these regulations is clear: to protect human health against the hazards presented by the use of materials containing unwanted substances.

To illustrate how these regulations impact on the choice and use of solvents in food use, I have chosen as a prime model of rules the Code of Federal Regulations of the United States. The reason for this choice is that although the Code of Federal Regulations has no legal status outside the United States, it is nevertheless very often referred to in other parts of the World.

But reviewing all the aspects of the Code of Federal Regulations dealing with solvents in food-related end-uses would be an immense task in itself. I will therefore limit this presentation to the study of the use of solvents on one type of foodstuff at various stages of its life.

That foodstuff could be an oilseed, for instance.

The first possible case of contact with a hydrocarbon solvent is seed dressing. Seed dressing is commonly practised to preserve the seed from attack and destruction by parasites. Technically, this operation is very similar to the spraying of pesticides for crop protection. In both cases, the role of the solvent is to facilitate the use and to enhance the efficiency of the active ingredients, and also to reduce the cost. Pesticides formulations, legitimately enough, are strictly regulated.

In the United States, the Environmental Protection Agency regulates the type of hydrocarbons that may be used in pesticide formulations. Part 40.CFR.180 of the Code of Federal Regulations lists the "Tolerances and Exemption from Tolerances for Pesticide Chemicals in or on raw Agricultural Commodities". In the general case, each component of a pesticide formulation used on crops, has, associated with it, a tolerance figure which represents the maximum amount that can be left on the crop. If, however, it is judged on the basis of suitable evidence that a particular chemical is harmless to certain plants, that chemical can be exempted from the need to have a tolerance when it is used on those plants. In practice this exemption provides the indication to the applicator and to the formulator that the solvent portion of the formulation presents a low phytotoxicity hazard.

Table 1 gives the boiling range and aromatic content of four grades of heavy aromatic solvents made by Essochem. Their molecular structure gives them a high solvency power as evidenced by the low mixed aniline point shown. Their controlled volatility characteristics due to narrow boiling ranges render these grades particularly attractive in pesticide formulations.

Table 2 presents the exemptions that were granted by the Environmental Protection Agency after examination of these solvents. Due to its special classification by the EPA, SOLVESSO 100 may be used without restrictions on growing crops and on grain under storage. Besides, it may also be used as inert- or occasionally active- ingredient in formulations applied to animals.

TABLE 1

ESSOCHEM HEAVY AROMATIC SOLVENTS			
	TYPICAL BOILING RANGE (°C)	MIXED ANILINE POINT (°C)	AROMATIC CONTENT (wt %)
SOLVESSO 100	165 - 179	14	99
SOLVESSO 150	190 - 209	16	99
SOLVESSO 200	226 - 286	15	98
HAN	193 - 282	20	84

TABLE 2

ESSOCHEM HEAVY AROMATIC SOLVENTS	
GRADE	EXEMPTED FROM TOLERANCE ON:
SOLVESSO 100	- Growing crops - Grain under storage - Animals
SOLVESSO 150	- Growing crops
SOLVESSO 200	- Growing crops
HAN	- Growing crops

If we continue the oilseed example, the next steps are likely to include solvent extraction for the recovery of the oil. Hexane is almost universally used as the solvent in this technology. The World Health Organization has issued guidelines concerning the quality of hexane that should be used. The guidelines comprize two series of tests: one for the quick identification of the material. As shown in Table 3, this series of tests includes: solubility in water, density and refractive index. The other series is for measuring the purity of the solvent. The next table lists the tests that must be carried out in the first column. Of particular importance in connection with food are aromatic content - essentially benzene - and the presence of polynuclear aromatics determined by ultra-violet absorption. In column 2 we have the maximum levels recommended by the World Health Organization, and in column 3 we see the typical values obtained on EXSOL hexane. The Code of Federal Regulations does not regulate residual hexane in edible oils in general but in some cases it is fixed. For instance in spice oleoresins where the limit is 25 parts per million.

TABLE 3

GUIDELINES OF THE WORLD HEALTH ORGANIZATION

HEXANE

IDENTIFICATION TESTS

Solubility in Water

Density

Refractive Index

TABLE 4

GUIDELINES OF THE WORLD HEALTH ORGANIZATION		
HEXANE PURITY TESTS	W.H.O.	EXSOL HEXANE
Residue on Evaporation, max.	0.0005 %	0.0005 %
Reaction of Residue	neutral	neutral
Distillation Range (°C)	64-70	65-70
Aromatic Hydrocarbons (%)	0.2	0.01
Sulfur, max. (mg/kg)	5	< 1
Lead, max (mg/kg)	1	<< 1
UV Absorbance: 280-289 nm	0.15	0.02
290-299 nm	0.13	0.01
300-359 nm	0.08	0.005
360-400 nm	0.02	0.000

Let us now carry our example another step. If we assume that the extracted meal is treated further to become food for human consumption, it can come in contact with hydrocarbon solvents in different ways, depending on the type of processing that it must undergo. It is probably appropriate at this point to recall that the FDA makes a distinction between uses where there is a direct contact between the solvent and the food and those where the contact is only indirect. The FDA defines the categories of solvents that are permitted in each case.

Table 5 presents some examples of direct food contacting. They are taken from the food processing section. Direct contact is indeed practically always required in this type of application.

As shown in Table 6, the only categories of solvents permitted in these cases are Synthetic Isoparaffinic Petroleum Hydrocarbons and Odourless Light Petroleum Hydrocarbons. Essochem grades meeting the specifications set for these categories are shown in the last column.

TABLE 5

EXAMPLES OF DIRECT CONTACT IN FOOD PROCESSING

OPERATION

Manufacture of Vinegar

Beet Sugar Processing

Cleaning of Vegetables

Citric Acid Manufacture

Insecticide Formulations
Used on Processed Foods

TABLE 6

EXAMPLES OF DIRECT CONTACT IN FOOD PROCESSING

OPERATION	SOLVENT CATEGORY	EXAMPLES
Manufacture of Vinegar	Synthetic Isoparaffinic Petroleum Hydrocarbons	ISOPAR Solvents
Beet Sugar Processing		
Cleaning of Vegetables	and	
Citric Acid Manufacture		
Insecticide Formulations Used on Processed Foods	Odourless Light Petroleum Hydrocarbons	EXSOL D60

Once the food has been processed, it must be packed: vegetable oil will be bottled or canned, solid or semi-solid food derived from the seed will be disposed in appropriate containers.

FDA regulations impact also on the choice of the solvents used in the manufacture of packaging materials for food: paper, paperboard, aluminium and plastics.

Hydrocarbon solvents are indeed used in the packaging industry for various purposes: as components of adhesives, in the formulation of additives to paper and paperboard, as production aids, etc. In all cases, the solvents used in these applications fall, according to the FDA, in the category of "Indirect Food Additives".

Subpart B of 21.CFR.176 deals with components of paper and paperboard used for the wrapping of food. The FDA regulations distinguish between hydrocarbons permitted for aqueous and fatty foods on one hand, and hydrocarbons for dry food on the other hand. For the wrapping of aqueous and of fatty foods, due to the fact that conditions for migration of components from the container into the food are aggravated by the presence of water or fat, only White Mineral Oil is permitted. For dry foods, only Synthetic Isoparaffinic Petroleum Hydrocarbons, Odourless Light Petroleum Hydrocarbons and toluene are permitted. These points are presented in Table 7.

TABLE 7

HYDROCARBON SOLVENTS PERMITTED AS COMPONENTS OF PAPER AND PAPERBOARD IN CONTACT WITH FOOD

Synthetic Isoparaffinic Petroleum Hydrocarbons
(ISOPAR grades)

Odourless Light Petroleum Hydrocarbons
(EXSOL D 40, EXSOL D 60)

Toluene
(SOLVESSO Toluene)

Aluminium foil is another material used widely to wrap food nowadays. During the rolling of the aluminium, lubricants are used to reduce friction, decrease energy consumption and produce a better finish of the surface. As lubricant left on the metal could come in contact with food - indirect food contacting - standards have been set up by the FDA for the quality of materials permitted in lubricant formulations.

Various classes of hydrocarbons are allowed in this application. A certain number of grades of Essochem Europe meet the most stringent specifications set for this end-use as shown on Table 8. It is noteworthy that less refined materials classified as Light Petroleum Hydrocarbons are also permitted provided that the ultra violet absorbance of the residue left on the metal after processing is measured and found to be within specifications.

TABLE 8

INDIRECT FOOD ADDITIVES

SURFACE LUBRICANTS USED IN THE MANUFACTURE OF METALLIC ARTICLES

Synthetic Isoparaffins	ISOPAR Grades
Mineral Oil	EXSOL D 100
Technical White Mineral Oil	EXSOL D 80 - VARSOL 75
	EXSOL D 40 - EXSOL D 60
Light Petroleum Hydrocarbons	VARSOL 60 - VARSOL 40

Plastics is another material that is widely used for packaging. Again its use for food wrapping is severely regulated. To take just one example: the packaging of meat and poultry in the United States is under the jurisdiction of the U.S. Department of Agriculture. The nature of the components entering the manufacture and the composition of plastics for this end-use are examined by this Department. When the component is recognized safe, a letter of acceptance is issued by the USDA to the manufacturer who may then utilize that chemical in his formulations. The USDA relies largely for his evaluations on the knowledge accumulated by the FDA.

While in the United States the Code of Federal Regulations lays down requirements for a comprehensive range of materials to be used in direct and indirect contact with food, a similar regionwide guideline is not available yet in Europe. However, for a number of years, many European countries have developed their own national rules to cope with the growing use of solvents in food-related uses.

For instance, in the U.K., the Ministry of Agriculture, Fisheries and Food has come up with a series of measures to control the use of solvents in pesticides.

In Germany, the Bundesgesundheitsambt has enacted laws governing the use of materials to be used in food packaging. It is worth saying a few more words about the approach taken in this case as it is rather unusual: independent laboratorieshave been authorized by the Government to test materials and to issue certificates of conformity for those that meet the requirements of the Law.

Table 9 shows the kind of document that is issued when a solvent meets the requirements set up for a particular application. In this case it is certified that the five solvents:
- SOLVESSO Toluene
- SOLVESSO Xylene
- SOLVESSO 100
- SOLVESSO 150
- VARSOL 145 / 200

may be used in additives for paper and plastics packaging in direct contact with dry, wet and fatty foods.

We could multiply examples like this. Every year indeed more regulations are issued and enforced in one country or the other. Signs are that the pace is not going to slow down in the years to come either: the EEC is also involved in legislation concerning solvents:

- Directive on the approximation of the Laws, regulations and administrative provisions relating to the classification, packaging and labelling of dangerous substances.

- Directive on the approximation of Member States' laws, regulations and administrative provisions relating to the classification, packaging and labelling of dangerous preparations (solvents)

- Directive on the approximation of the laws of the Member States concerning the preservatives authorized for use in foodstuffs intended for human consumption, etc.

Harmonization of the legislation will thus take place in the future, at least to a certain degree. The level of complexity of the resulting regulations, however, can be expected to remain rather high, due to the specificity of the applications. Therefore potential solvents users will continue to seek information and advice from the solvents suppliers, and the responsible suppliers will have to make the effort to be able to provide that guidance.

The ESSO CHEMICALS emblem and the terms: ISOPAR, VARSOL, SOLVESSO, EXSOL, HAN, are Trademarks of Exxon Corporation.

A COMPUTER PROGRAM FOR THE LABELLING
OF DANGEROUS SUBSTANCES

G. Jacobs, M. Martens and L. Hulsen

Institute for Hygiene and Epidemiology
J. Wytsmanstraat 14, 1050 Brussels.

1. INTRODUCTION

As the 79/831/EEC directive on the classification, packaging and labelling of dangerous substances will be soon implemented in all member states of the EEC, the manufacturers and importers of dangerous chemicals shall be obliged to label those substances which are not included in the annex I of the 67/548/EEC directive. (original directive on dangerous substances).
Since it is the aim of this directive to prevent commercial barriers within the EEC due to different labelling of dangerous chemicals, precise criteria of labelling should be set forward. The commission of the EEC made a great effort to establish a labelling guide as a part of the VI th annex to the 79/831/EEC directive. To date no full agreement has been obtained among the member states on some of the criteria. The most important amongst them are the criteria in relation to carcinogenic, mutagenic and teratogenic compounds. It is hoped that a compromise will be obtained early this year.
Our Institute is actually involved in the labelling of organic solvents not being included in the annex I of the 67/548/EEC directive. 200 solvents have already been examined and proposals for labelling have been made. In doing so we realised soon that it is very difficult to label in a reproducible manner even when using the EEC drafts of the labelling guide. Labelling proposals differed from person to person and even changed for the same individual in function of time. The greatest difficulty was to consider for each substance all aspects (physico-chemical, chemical, toxicity,

safety, disposal etc.) at the same time. This of course
on the basis of a as complete as possible set of data.
To guarantee an objective and reproducible labelling and
hence classification of dangerous substances, we developed a computer programme. The programme, which is written in
IBM-basic is based on the actual position of the EEC working group on labelling and on our own practical experience.
For the criteria on which no agreement has been obtained at
the EEC, the Belgian position is taken into account.
The symbols, risk phrases and safety phrases as they are
used in the computer programme are explained in the tables
1, 2 and 3.

2. DESCRIPTION OF THE COMPUTER PROGRAM

The programme is conceived in the form of an interrogation procedure (tables 4 and 5, these tables don't represent the flow diagram, but are a summary of the questions
that appear on the computer screen).
First, questions are asked on the physical form and on
physico-chemical data such as molecular weight, boiling
point, vapour pressure and flash point (not shown in table 4). The figures entered are stored in the computer memory for further treatment.
The physical form of the substance (solid, liquid, gas, liquified gas) is required for several reasons such as (a) to
exclude gases from inquiries on physico-chemical data (molecular weight, boiling point, vapour pressure and flash
point) ; (b) to exclude gases from the question "substance
may generate gas and burst the container" (table 4-H) ;
(c) to separate gases from liquids and solids to examine
the flammability data (table 4-F) ; (d) to exclude liquids
and gases from questions on flammability data meant for solids (table 4-G) ; (e) if solids are found to be toxic S22
is assigned, if toxic liquids of low volatility are concerned S23 is proposed.
The molecular weight (liquids and solids) is needed for the
calculation of 25 % of the concentration in the air at saturation. This value is compared with the threshold limit value or with the no effect level (table 5-J) for the assignment of S38.
The boiling point (liquids and solids) at 101080 Pa is required to assign (a) the symbol FF to substances with a
boiling point below 35 °C and a flash point below 0° C
(table 4-F) ; (b) S3 to substances with a boiling point below 41 °C ; (c) S38 to substances which are found to be
harmful by inhalation and have a boiling point below 100 °C;

(d) S23 to liquids which are found to be toxic by inhalation and have a boiling point beyond 100 °C.
The vapour pressure at 20-25 °C is required for solids and liquids to assign (a) S38 to substances which are found to be harmfull by inhalation and have a vapour pressure beyond 500 Pa ; (b) S23 to liquids which are found to be toxic by inhalation and have a vapour pressure below 500 Pa. The vapour pressure is also needed for the calculation of the concentration in the air at saturation (for the assignment of S38).
The flash point obtained with the closed cup method is used for the attribution of FF, F or R 10 (table 4-F).
If no data are available on the boiling point and on the vapour pressure, the computer sets these values automatically equal to zero. If no data are available on the flash point the ceiling value (999999) is taken.
As far as the toxicity of the compound is concerned, first data on acute toxicity are asked (table 5-I). These data are stored in the computer memory to be altered afterwards according to the degree of hazard of other toxic effects (table 5-N,O,P). The original acute toxicity data can only be upgraded due to other toxic effects.
If corrosion of the skin is observed (man or rabbit) no more questions are asked on inflammation or sensitisation of the skin and on corrosion or inflammation of the eye (table 5-K). The answer to the question "Does substance penetrate the skin" (table 5-K) is stored. It will be used afterwards to investigate whether or not subcutaneous data on toxic effects other than acute (table 5-N,O,P) may be used to consider the substance as hazardous via the percutaneous route.
It is important to mention that, until now, no full agreement has been obtained on the interpretation of the toxic effects displayed in table 5-N,O,P. Therefore the assignment of the symbols and the R-and S-phrases is based on the Belgian position in question (02/1982). However, it has been convened upon that the route of administration relevant for the toxic effect should be indicated. This is done by a subroutine called "subr. administration". This procedure investigates for each effect the responsible route of administration. If, for example, a human carcinogen only acts through inhalation, the LC50 value stored is set to 1 (if LC50 ihl > 500 mg/m^3), according to the hazard level of this effect, namely TT (very toxic). As another example we may consider an irreversible (other than genotoxic) effect caused by repeated or prolonged skin exposure. In this case the percutaneous LD50 value stored is set to 100 (if LD50 skn>400 mg/kg) according to the hazard level of this effect, namely T (toxic).

This subroutine also takes into consideration the routes of administration which are pharmacokinetically the closest to the three standard routes (oral, percutaneous and by inhalation). For example, many data on teratogenicity of chemicals have been obtained via the intraperitoneal route. It would be erroneous to neglect these in case no oral data are available. Therefore the intraperitoneal route is used as a basis to propose risk phrases for oral intake. The same type of reasoning is applied if no data are available on percutaneous toxicity. In this case, subcutaneous data are considered on the condition that it is known that the substance can be absorbed through the skin (tabel 5-K).

As far as mutagenic substances are concerned a supplementary question is asked on the necessity of a bioactivation system to obtain the mutagenic effect. On the basis of this answer it can be decided whether or not "first-pass" free routes of administration should be considered. In this case R-phrases of the Xn-level are assigned.

When the interrogation procedure is terminated, the data stored are treated. First it is examined in what circumstances the S38 should be proposed (already discussed before). Afterwards, the acute toxicity data (whether or not altered according to non-acute toxicity effects) are compared to the hazard levels of acute toxicity set by the EEC (table 6). By doing so the corresponding symbols (TT, T, Xn), R-phrases (20,21,22,23,24,25,26,27,28) and S-phrases (1,36,37,44,45) are assigned, in addition to the set of phrases already mentioned in tables 4 and 5. Before the final proposition of the label, the computer asks the operator to add, if necessary, one or more of the following S-phrases : S4, S8, S15, S21, S40, S42. These S-phrases were not considered in the computer programme for want of precise criteria.

When the procedure is finished a label is proposed. The computer will present the symbol(s) and all the R-and S-phrases which reflect the hazard of the substance and the safety measures to be taken. It will remain the responsibility of the competent EEC working group to reduce each set of phrases to four. This because no precise criteria exist on the priorities for R-and S-phrases.

The label presented is devoid of phrases ment for use in the home. However, together with this label a set of phrases is listed which should be considered if the substance is to be used in the home :
S2, S13, S20, S45 if the symbol is TT,T,CC,C or Xn ;
S19 instead of R32 or R31 ; S25 instead of S39 if R41,
S32 instead of S38.

3. CONCLUDING REMARKS

It is important to mention that (a) the complete flow diagram (22 pages) is not presented in this paper and (b) there are parts in the computer programme for which, until now, no European consensus has been obtained. Although the programme has been thoroughly tested on more than 100 substances, we feel that we must await a full agreement on all aspects of labelling.
If this is obtained the programme will be adapted and published in full as an extensive report of our institute. Much more attention shall paid to what we understand by "human carcinogens","putative carcinogens", "irreversible effects other than genotoxic","reversible effects" etc. Lots of examples will be shown to demonstrate the performance of the programme and the completeness of the data needed to answer properly to the questions asked. Even when a standardized strategy like a computer programme renders labelling more uniform, the quality of the set of data on physico-chemistry and toxicity will remain the most important basis for the correct labelling of dangerous substances.

ACKNOWLEDGEMENTS

The authors wish to thank Prof. Dr. A. Lafontaine, director of the institute for his interest in their work. They also are grateful to I. Van Droogenbroeck and L. Coopman for the preparation of the transparencies and the type-writing.

TABLE 1

Symbols used in the computer programme

EE*	,	exploding bomb	: very explosive
E	,	exploding bomb	: explosive
O	,	flame over a circle	: oxidizing
FF*	,	flame	: extremely flammable
F	,	flame	: highly flammable
TT*	,	skull and cross-bones	: very toxic
T	,	skull and cross-bones	: toxic
Xn	,	St Andrew's cross	: harmful
CC*	,	showing damaging effect of an acid	: very corrosive
C	,	showing damaging effect of an acid	: corrosive
Xi	,	St Andrew's cross	: irritant

* These symbols are not recognized by the EEC, they are used in the computer programme to differentiate between two levels of hazard without considering the risk phrases.

TABLE 2

R- phrases, situation 03/1982

R 1 Explosive when dry
R 2 Risk of explosion by shock, friction, fire or other sources of ignition.
R 3 Extreme risk of explosion by shock, friction, fire or other sources of ignition.
R 4 Forms very sensitive explosive metallic compounds.
R 5 Heating may cause an explosive.
R 6 Explosive with or without contact with air.
R 7 May cause fire.
R 8 Contact with combustible material may cause fire.
R 9 Explosive when mixed with combustible material.
R 10 Flammable
R 11 Highly flammable
R 12 Extremely flammable.
R 13 Extremely flammable liquefied gas.
R 14 Reacts violently with water.
R 15 Contact with water liberates highly flammable gases.
R 16 Explosive when mixed with oxidizing substances.
R 17 Spontaneously flammable in air.
R 18 In use, may form flammable/explosive vapour-air mixture.
R 19 May form explosive peroxides.
R 20 Harmful by inhalation.
R 21 Harmful in contact with skin.
R 22 Harmful if swallowed.
R 23 Toxic by inhalation.
R 24 Toxic in contact with skin.
R 25 Toxic if swallowed.
R 26 Very toxic by inhalation
R 27 Very toxic in contact with skin.
R 28 Very toxic if swallowed.
R 29 Contact with water liberates toxic gas.
R 30 Can become highly flammable in use.
R 31 Contact with acids liberates toxic gas.

TABLE 2, cont'd

R-phrases, situation 03/1982.

R 32 Contact with acids liberates very toxic gas.
R 33 Danger of cumulative effects.
R 34 Causes burns.
R 35 Causes severe burns.
R 36 Irritating to eyes.
R 37 Irritating to respiratory system.
R 38 Irritating to skin.
R 39 Danger of very serious irreversible effects.
R 40 Possible risk of irreversible effects.
R 41 Risk of serious damage to the eye.
R 42 May cause sensitization by inhalation.
R 43 May cause sensitization by skin contact.
R 49 Carcinogen/Known to cause cancer.
R 50 Suspected carcinogen/May cause cancer.
R ? Risk of explosion if heated under confinement.

R 14/15	Reacts violently with water liberating highly flammable gases.
R 15/29	Contact with water liberates toxic, highly flammable gas.
R 20/21	Harmful by inhalation and in contact with skin.
R 21/22	Harmful in contact with skin and if swallowed.
R 20/22	Harmful by inhalation and if swallowed.
R 20/21/22	Harmful by inhalation, in contact with skin and if swallowed.
R 23/24	Toxic by inhalation and in contact with skin.
R 24/25	Toxic by contact with skin and if swallowed.
R 23/25	Toxic by inhalation and if swallowed.
R 23/24/25	Toxic by inhalation, in contact with skin and if swallowed.
R 26/27	Very toxic by inhalation and in contact with skin.
R 27/28	Very toxic in contact with skin and if swallowed.
R 26/28	Very toxic by inhalation and if swallowed.
R 26/27/28	Very toxic by inhalation, in contact with skin and if swallowed.
R 36/37	Irritating to eyes and respiratory system.
R 37/38	Irritating to respiratory system and skin.
R 36/38	Irritating to eyes and skin.

TABLE 2, cont'd

R-phrases, situation 03/1982.

R 36/37/38	Irritating to eyes, respiratory system and skin.
R 42/43	May cause sensitization by inhalation and skin contact.

TABLE 3

S-phrases, situation 03/1982

S 1   Keep locked up.
S 2   Keep out of reach of children.
S 3   Keep in a cool place.
S 4   Keep away from living quarters.
S 4 -bis Keep at temperature not exceeding...°C (to be indicated by the manufacturer).
S 5   Keep contents under ... (appropriate liquid to be specified by the manufacturer).
S 6   Keep under ... (inert gas to be specified by the manufacturer).
S 7   Keep container tightly closed.
S 8   Keep container dry.
S 9   Keep container in a well-ventilated place.
S 10 Keep wetted with... (to be indicated by the manufacturer).
S 11 Keep in the original container.
S 12 Do not keep the container sealed.
S 13 Keep away from food, drinks and animal feeding stuffs.
S 14 Keep away from...(incompatible materials to be indicated by the manufacturer).
S 15 Keep away from heat.
S 16 Keep away sources of ignition - No smoking.
S 17 Keep away from combustible material.
S 18 Handle and open container with care.
S 19 Do not mix with... (to be specified by the manufacturer).
S 20 When using do not eat or drink.
S 21 When using do not smoke.

TABLE 3, cont'd

S-phrases, situation 03/1982.

S 22 Do not breathe dust.
S 23 Do not breathe gas/fumes/vapour/spray.
S 24 Avoid contact with skin.
S 25 Avoid contact with eyes.
S 26 In case of contact with eyes, rinse immediately with plenty of water and seek medical advice.
S 27 Take off immediately all contaminated clothing.
S 28 After contact with skin, wash immediately with plenty of... (to be specified by the manufacturer).
S 29 Do not empty into drains.
S 30 Never add water to this product.
S 32 Use only in well ventilated areas.
S 33 Take precautionary measures against static discharges.
S 34 Avoid shock and friction.
S 35 This material and its container must be disposed of in a safe way.
S 36 Wear suitable protective clothing.
S 37 Wear suitable gloves.
S 38 In case of insufficient ventilation, wear suitable, respiratory equipment.
S 39 Wear eye/face protection.
S 40 To clean the floor and all objects contaminated by this material, use ... (to be specified by the manufacturer).
S 41 In case of fire and/or explosion do not breathe fumes.
S 42 During fumigation/spraying wear suitable respiratory equipment.
S 43 In case of fire, use... (indicate in the space precise type of fire-fighting equipment. If water increases the risk, add - Never use water).
S 44 If you feel unwell, seek medical advice (show the label where possible).
S 45 In case of accident or if you feel unwell, seek medical advice immediately (show the lable where possible).
S 46 Pregnant women should not be exposed to this product.

S 1/2 Keep locked up and out of reach of children.
S 3/9 Keep in a cool, well-ventilated place.
S 7/9 Keep container tightly closed and in a well-ventilated place.
S 7/8 Keep container tightly closed and dry.
S 20/21 When used do not eat, drink or smoke.
S 24/25 Avoid contact with skin and eyes.

TABLE 3, cont'd

S-phrases, situation 03/1982.

S 36/37	Wear suitable protective clothing and gloves.
S 36/39	Wear suitable protective clothing and eye/face protection.
S 37/39	Wear suitable gloves and eye/face protection.
S 36/37/39	Wear suitable protective clothing, gloves and eye/face protection.

TABLE 4

Physico-chemical data

Questions asked by the computer		Symbols and phrases assigned according to the answers given (Y/N/U)		
		Symbols	R	S
A. Reactivity data	- May cause fire. - Violent reaction with water. - Violent reaction with oxidizing materials.		7,14	7,14,30 43

B. Explosivity data in relation to the compound itself	- Explosive by chock, friction, ignition
	  . lability comparable with Pb-azide
	  . is compound an organic peroxide
	- Explosive by heat other than ignition
	- Explosive at ambient temperature, even without oxygen
	- Only labile when dry
		EE,E	1,2,3 5,6	3,7,9,15 17,34,35, 37,39

TABLE 4, cont'd

Physico-chemical data

C. Explosivity data in relation to the reaction of the compound with other compounds	– Reacts explosively with other substances • oxidizing materials • combustible materials • other materials	O	9,16	14,17
D. Explosivity data in relation to reaction products of the compound	– Can form explosive compounds • metal derivatives • peroxides • other labile products		4,19	14,18
E. Oxidizing properties	– Is substance an organic peroxide – Fire in contact with combustible materials	O	8	3,7,9,17 37,39
F. Flammability data	Computer programme compares the data entered for boiling point, physical form and flash point with the criteria set by the EEC	FF,F	10,11,12	9,16
	– Flammable gas	F	12,13	9,16
	– Ignites in contact with air	F	17	6

TABLE 4, cont'd

Physicochemical data

		F		
G. Flammability data in relation to solids	- Ignites spontaneously in contact with air - Ignites immediately in contact with heat - Forms highly flammable gas in contact with water		11,15 17	5,7,16,43
H. Miscellaneous data	- Substance may generate gas and burst the container			12
	- Is the liquid miscible with water			29
	- Does gas, liquid or solid in powder form accumulate electrostatic charges			33

TABLE 5

Toxicity data

Questions asked by the computer		Symbols and phrases assigned according to the answer given (Y/N/U)		
		Symbol	R	S
I. Acute toxicity data	– LD50 orl, rat............mg/kg. – LD50 skn, rat or rbt........mg/kg. – LC50 ihl, rat............mg/m³ in ...hrs	figures entered are saved		
J. – Threshold limit value ....mg/m³ – No effect level ....mg/m³				

K. Skin effect data	– Corrosion (necrosis, tissue destruction)
	  . observed in animal after 3 min contact
	  . observed in man as severe burns
	– Inflammation (redness, swelling)
	– Sensitization
		CC,C
		Xi
			34,35,36
			38,41,43
				1,24,25,
				26,27,28,
				36,37,38,
				39

	– Does substance penetrate the skin	answer is saved		
L. Eye effect data	– Corrosion (necrosis, tissue destruction) – Inflammation (redness, swelling)	Xi	36,41	25,26,39

TABLE 5, cont'd

Toxicity data

M. Effects on respiratory organs	- Sensitisation - Irritation	Xn,Xi	37,42	38
N. Carcinogencity and mutagencity data	- Known human carcinogen - Putative human carcinogen - Questionable carcinogen - Tumor promotor or co-carcinogen - Mutagen . is a bioactivation system necessary for mutagenic or clastogenic action.	TT,T Xn	40,49 50	38
O. Teratogenicity data	- Teratogenic in humans - Teratogenic in animals	T,Xn	39,40	46

P. Other toxicity data	- Irreversible effects, other than geno-toxic effects
	  . caused by a single exposure
	  . caused by repeated or prolonged exposure
	- Long term effects, other than mentioned before
	  . severe effects
	  . moderate effects
	- Evidence for cumulative effects.
		TT,T
		Xn
			33,39	

TABLE 5, cont'd

Toxicity data

Q. Formation of toxic gases	– With water – With acids · are the gases formed very toxic – On combustion or explosion.	29,31, 32	7,30, 41

TABLE 6

*Classification of dangerous substances on the basis of acute toxicity data.
(79/831/EEC).*

Category	oral LD50, rat (mg/kg)	percutaneous LD50, rat or rabbit (mg/kg)	inhalation LC50 (mg/L/4h)
Very toxic	⩽25	⩽50	⩽0.5
Toxic	>25, ⩽200	>50, ⩽400	>0.5, ⩽2
Harmful	>200, ⩽2000	>400, ⩽2000	>2, ⩽20

EUROPEAN SYSTEMS FOR THE LABELLING OF SOLVENTS

S.G. Luxon

Safety & Health Consultants
47 Weymouth Street
London W.1.

During the last 25 years there has been an increasing awareness of the hazards associated with chemicals of all kinds. Simultaneously there has been a demand, extending more widely, for open consideration of the risks which may be involved in activities where such chemicals are present. One of the outcomes of these pressures, so vividly presented by the media, has been a demand for the more positive identification of hazardous materials coupled with information on the risks which they pose during use.

Recent surveys in the U.S. have shown that 1 in every 4 workers is exposed to some 400 regulated hazardous chemicals and almost 1 in 10 to possible carcinogenic substances. Some 86,000 trade name chemicals are used and in one third, at least one other trade name product was present. In such trade name products neither the worker nor the employer is in a position to recognise the hazard so the demand identified above for a labelling system appears to be more than justified.

Clearly the task of meeting such demands is one of great magnitude and is perhaps best dealt with by establishing a system of priorities triggered either by the scale of use or, by the severity of the hazard, or by the introduction of a new substance for the first time.

Labelling which forms perhaps the simplest universal way of alerting those who might be exposed to deleterious effects was, in the earliest deliberations, associated with the first of these priorities i.e. the scale of use of the particular chemical.

In the 1950's when the need for action was first recognised, discussions were taking place on possible general broad classification systems under the aegis of a number of international organisations - the United Nations, the International Labour Organisation and the Council of Europe being perhaps the more important. The U.N. groups were primarily concerned with transport which perhaps has the widest international implications while the I.L.O. and C.O.E. were concerned principally with hazards (which were of greater national importance) to the user. It was recognised at this early stage that there was a need for unification but it proved impossible to agree on a single system in each of these fields let along in both.

The I.L.O. in 1959 adopted symbols of uniform colour in a square or rectangular format which were subsequently incorporated into labelling systems for users although even those represented a compromise on the symbol for corrosive materials incorporating as it does two indications i.e. the corroded hand and the corroded metal bar.

On the other hand the U.N. transport system had adopted a diamond format with different colours to indicate the type of hazard, and utilised similar symbols in the upper half, with key words in the lower half.

In view of the difficulties which had appeared in the I.L.O. deliberations, it had become apparent to all concerned that any attempt to formulate a truly international system for user labelling would be an impossibly protracted task. The C.O.E. accordingly took on board the more limited objectives of developing recommendations for a European system within the so-called "Partial Agreement" countries - broadly the E.E.C. member states and Austria, for the labelling of "pure" substances.

These deliberations resulted in the publication of the so-called "Yellow Books" entitled "Dangerous Chemical Substances and Proposals concerning their labelling". Following the formation of the E.E.C., the Commission proposed that this system should be adopted by member states under Art. 100 of the Treaty of Rome to facilitate the free trade in such products between member states. The Basic Directive of 1967 and subsequent admending Directives have been the result.

Essentially this system comprised a warning symbol which would be universally recognised coupled with a key word and risk and safety phrases. These phrases were standardised and numerically numbered so that they could be published and cross referenced in the five languages then recognised as relevant.

Originally, with relatively few substances under consideration it was possible to decide on labelling in respect of each individual substance in the light of the practical experience of members of the group.

It soon became clear, however, that such a course of action would in the longer term result in anomalies particularly where one delegation had in the past experienced severe repercussions in respect of a particular chemical which might be due to local circumstances rather than an intrinsically high hazard. Accordingly efforts were made to rationalise the system as far as this was possible given the state of the art at the time.

The group accordingly turned its attention to the general classification of hazardous substances and agreed certain broad parameters for Toxic, Harmful and Flammable substances but could only make more general recommendations for the classification of explosive, corrosive and oxidising substances.

For Toxic and Harmful substances the oral LD_{50} rat was taken as the determining factor and general guidelines were agreed. A simplistic approach such as this based on acute toxicity in animals alone cannot adequately take account of the many long term effects which are difficult to quantify. The overall assessment of toxicity must be related to experience gained from epidemiological studies, human experience in case of accidental poisoning and possible sensitisation, carcinogenicity and mutagenic effects. Each substance was therefore considered individually to determine whether or not the simple classification was adequate.

A great deal of discussion also took place on the alternative possibility of using dilution volumes and volatility where applicable, but as no agreement could be reached on this approach it was abandoned. This was unfortunate as subsequent difficulties in respect of the labelling of solvents has demonstrated.

Gradually a more orderly system was built up in which not only the symbols but also the risk phrases were allocated to substances according to certain well defined rules: subject of course to any over-riding special hazards in a particular case which would amply justify a change from this general classification.

As they apply to pure solvents and pure constituents of solvents these general rules can be summarised as follows:

SYMBOL	MEDIAN LD_{50} Oral Rat	MEDIAN LD_{50} Percutaneous Rat/Rabbit	MEDIAN LC_{50} Inhalation Rat (4hr)
(Very Toxic	$\leqslant 25$mg/kg	$\leqslant 50$mg/kg	$\leqslant 0.5$mg/l
*(Toxic	> 25 to 200mg/kg	> 50-400mg/kg	> 0.5-2mg/l
(Harmful	> 200 to 2000mg/kg	> 400-2000mg/kg	> 2-20mg/l

*Note the same symbol is used for Toxic and Very Toxic.

PHRASES TOXIC EFFECTS

Risk phrases may be regarded as further defining the risk category indicated by the symbol. It is therefore useful to group them under each symbol type.

SYMBOL	RISK PHRASE	NUMBER	PARAMETERS
Very Toxic	Very Toxic (By Inhalation	R26 LC_{50}	$\leqslant 0.5$mg/L(4hr)
	(By Skin Contact	R27 LD_{50}	$\leqslant 50$mg/kg dermal
	(If Swallowed	R28 LD_{50}	$\leqslant 25$mg/kg oral
Toxic	Toxic (By Inhalation	R23 LC_{50}	0.5-2mg/L(4hr)
	(By Skin Contact	R24 LD_{50}	50-400mg/kg dermal
	(If Swallowed	R25 LD_{50}	25-200mg/kg oral
Harmful	Harmful (By Inhalation	R20 LC_{50}	2-20mg/L(4hr)
	(By Skin Contact	R21 LD_{50}	400-2000mg/kg dermal
	(If Swallowed	R22 LD_{50}	200-2000mg/kg oral

Notes:
1. Phrases can be combined as appropriate.
2. Phrases R39, R48, R49 and R50 may have specific application to toxic substances.
3. Phrases R33, 40 and 42 may have specific application to harmful substances.

SYMBOLS CORROSIVE AND IRRITANT EFFECTS

Corrosive - substances which may, on contact with living tissue destroy them (see parameters below).

Risk Phrases	Parameters
Causes severe burns R35	When applied to the intact skin of an animal causes necrosis in 3 mins.
Causes burns R34	As R35 but in time up to 4 hrs.

LABELLING OF SOLVENTS

Irritant – non-corrosive substances which nevertheless through immediate, prolonged or repeated contact with skin or mucous membrane causes inflammation.

Risk Phrases	Parameters
Inhalation	
Irritating to the respiratory system R37	No parameters
May cause sensitisation by inhalation R42	

External (Skin & Eyes)	
Risk Phrases	Parameters
Irritating (to the eyes R36	Occular lesions as defined
(to the skin R38	Skin inflammation as defined

May cause sensitisation	
Risk Phrases	Parameters
By skin contact R43	Sensitisation as defined

Flammability	
Risk Phrase	Parameters
*(Extremely flammable R12	$FP < 0°C$ BP $35°C$
(Highly flammable R11	$FP \leqslant 21°C$

Notes

*Same symbol

For liquids having a $FP > 21°C \leqslant 55°C$ the word flammable is used without a symbol. R10.

It was also necessary to consider the priority of symbols and the maximum number to be used on any one label. For simplicity it was decided that only two symbols should be recommended – one for Toxic/Harmful/Corrosive properties and one for Flammable/Oxidising/Explosive properties.

The order of precedence was agreed as:

$$T \quad \text{over} \quad C \quad \text{over} \quad X_n \quad \text{or} \quad X_i$$

and

$$E \quad \text{over} \quad O \quad \text{over} \quad F$$

As with risk phrases, safety advice phrases were originally allocated on an adhoc basis and dealt with the detailed precautionary measures required. They therefore inevitably reflected the local circumstances of use and attempts were made to eliminate such detailed advice and the phrases were subsequently used to supplement and reinforce the risk phrases. Again a maximum number of four phrases was the aim making a total of eight phrases in all.

In addition the concept of combination phrases was developed so that a single sentence could combine two or three phrases with a consequent reduction in the words required to convey the same amount of information.

During the initial discussions, a strong plea was made for the opportunity to be taken in the amending Directive to bring together the transport and user systems. Unfortunately the exponents of neither system could agree to sufficient flexibility to enable meaningful discussions to take place. Under lying difficulties were pinpointed, such as the differing effects in the long and short term of certain toxic substances e.g. Benzene where in transport the principal risk is of fire whereas in use it is the long term irreversible toxic effects that are the more serious. Certain of the existing member states had also enacted the early 1967 Directive on user labelling. Thus the basic differences in shape and colour while appearing superficial had by then become enshrined in national thinking and in addition the classification of individual substances had followed a different pattern so that harmonisation was not merely a question of agreeing universal symbols and formats.

For all these reasons it proved impossible to change the widely held view that the basis of future action rested with two systems user and transport but the present UK proposal that for transport purposes containers of less than 250 litres should carry only the supply label will greatly simplify the problem as in practice transport labels will only be required on bulk loads.

We have up the present time been considering substances. The word 'substances' is used in this context in a particular way to denote pure or technically pure chemical elements and compounds which have not been mixed together in a deliberate way to provide a preparation. Substances and Preparations are dealt with in separate Directives. The Dangerous Substances Directive of 1976 sets out labelling requirements for each substance listed. The Preparations Directives on the other hand indicate how the hazards from these substances should be taken into account in deriving overall classification and labelling requirements from the mixture or preparations under consideration. Directives relating to "Solvents" have already been adopted by the Council of Ministers and a further amendment is under consideration. It should be noted that each individual component of these preparations appears in the basic substances Directive and it is from the classification in this basic Directive that the label for the preparation is derived.

The solvents directive proposes that to derive an overall classification for a mixture of pure substances each of those substances should be classified into three toxic sub classes and four harmful sub classes, each sub class having an index based on the intrinsic toxic properties and the volatility of the substance.

By multiplying these indices by the percentage by weight of each constituent the overall classification is determined.

This system is complex and unprecise and has not yet been finalised. Full details have been sent out in the recently published Consultative Document on the labelling of dangerous substances.

Reference has been made in a previous paper to possible alternative approaches and the subject is, I believe worthy of much consideration before a final solution is agreed.

Ideally a system which could combine the relevant characteristics of two or more pure solvents in a predetermined way so as to arrive at an overall classification for the mixture would have incalcuable advantages in that it would be possible without specific regulatory intervention (i.e. listing in cagegories) for the manufacture and correctly label any existing, new or modified preparation.

With the aid of the Computer Programme outlined in the next paper it should be possible to test a number of alternative procedures in the expectation that such a solution can be found.

In the time available I have outlined the present requirements and the thinking underlying them. A simple effective universal system easily understood and operated by all concerned is clearly the aim. Unfortunately, as we have seen the requirements are easy to state but in practice are difficult to achieve. There will undoubtedly be further refinements and extensions of the present proposals as knowledge of the practical applicationof the proposals and the intrinsic hazards of chemicals increases.

While there is a need to meet present requirements, let us not lose sight of the broader picture of which labelling

is only one small part - to alert all those who may come
into contact with solvents of the hazards associated with
their use and hence the need to provide anticipatory
precautionary measures to safeguard health and safety,
supplemented by training programmes designed to explain
more fully the hazards and their control.

SAFETY IN HANDLING/STORAGE OF SOLVENTS AT TERMINALS

Stephen A. Yuhas, Jr.

Exxon Chemical Company
Solvents Technology Division
P. O. Box 536
Linden, NJ 07036, USA

The handling and storage of solvents can be potentially hazardous because of their properties, including: flammability, high vapor pressure, physiological effects, ability to generate static electricity when flowing, and solvency effects on packings, gaskets, hoses.

Nevertheless, solvents can be handled and stored safely at terminals if one recognizes these potential hazards and takes suitable precautions. This paper deals with the safe practices for handling solvents, with a view toward minimizing the hazards. The checklist presented below may be a good starting point for review during a terminal safety audit. Although the checklist is not all inclusive, it can reveal potential problem areas.

PLANT/TERMINAL LAYOUT

☐ Allowance for minimum distances between --
  - Above-ground tanks and high-risk facilities such as, storage buildings, warehouse, office, garage, barrel filling, shipping/receiving, boiler house, electrical transformers -- 20 to 30 meters.
  - Above-ground tanks and railroad -- 30 meters.
  - Loading racks and high risk facilities -- 30 meters.
  - Loading racks and truck parking/waiting area -- 30 meters.

☐ Tanks and tank groupings surrounded by dikes.
  - Adequate containment capacity -- 100% of largest tank plus displacement of all other contained tanks.
  - Dikes are continuous. There must be no valves, drainpipes, or gates in the dikes. Only siphon drains are allowed. Stairs or platforms over dike walls allow access to tanks.

☐ Clear identification/color coding of tanks, pipelines,

valves.
- ☐ Tank field free of combustibles, such as weeds, grass, bushes, refuse.
- ☐ Ventilated drum storage area.
  - Drums are best stored in an open space or undercover, not in a closed building.
  - Drums are stored on their sides, and chocked to prevent rolling.
- ☐ Waste disposal provision. Approved waste solvent disposal methods should be available.
- ☐ Walkways, roads, parking areas adequate and well maintained.
- ☐ Adequate lighting for safe night operation.

STORAGE TANKS

- ☐ Clear identification of tank number and product at inlet and outlet valves.
- ☐ Vapor pressure/vapor loss control devices.
  - Pressure/vacuum vents for "breathing" during temperature changes, filling or withdrawing product, fire emergency, or overfilling.
  - Flame arrestors, screens on vents.
  - Emergency pressure release vent or weakened shell/roof seam.
  - Internal floating roof for low flash point solvents.
- ☐ Gauging provisions.
  - Automatic gauge accurate, reliable, and working.
  - Gauge hatches tightly closed to prevent escape of vapor. Check gasket condition.
  - Gauging well extending to bottom of tank to minimize possibility of electrostatic spark discharge.
- ☐ Discharge of inlet pipe located near the bottom of the tank to avoid splash filling and minimize turbulence, which could produce electrostatic charges.

ELECTRICAL EQUIPMENT AND ELECTROSTATIC CONCERNS

- ☐ Conformance of all electrical equipment to local electrical codes.
- ☐ Vapor-proof/explosion-proof lights and electrical equipment.
- ☐ Electrical <u>grounding</u> at all times of tanks, pipes, pumps, meters, loading racks, drum filling stations, etc. This can be accomplished through a common ground.
- ☐ Electrical <u>bonding</u> between tanks, pipes, pumps, meters, etc. Jumpers should be used across pipe flanges to

STORAGE OF SOLVENTS AT TERMINALS

ensure electrical continuity. Electrical continuity should be tested periodically.
- ☐ Grounding/bonding provisions for tankcars and tankwagons at loading racks.

LOADING/UNLOADING OPERATIONS

- ☐ Tankcars or tankwagons must first be electrically grounded.
- ☐ The vehicle must be electrically bonded to the loading rack.
- ☐ Pipelines and valves should be clearly identified at the loading rack.
- ☐ Avoid splash filling to minimize electrostatic charge formation.
  - Extend fill pipe to bottom of tankcar or tankwagon, or bottom load tankwagons.
  - Limit flow velocity to 1 meter/second until the inlet pipe opening is covered.
- ☐ Emergency "panic button" pump or valve shut off should be conveniently located. A "dead-man" switch is often used for this purpose.
- ☐ Allow time for electrostatic charge relaxation.
  - 5 minutes before gauging or sampling a filled tankcar or tankwagon.
  - 30 minutes before gauging large tanks or barges.
  - Filters can produce large electrostatic charges in flowing solvents. Piping design should allow a 30-second pipeline residence time between the filter and receiving tank for charge relaxation.
- ☐ Drum filling should be done outdoors, under shelter.
  - Electrically ground and bond drums during filling operations.
  - Use mechanical vapor exhaust to protect workers and minimize fire hazard.

PERSONNEL PROTECTION

- ☐ Protective equipment available and used -- resistant shoes, gloves, hard hats.
- ☐ Emergency showers and eye wash fountains.
- ☐ Emergency respiratory protection equipment.
- ☐ First aid equipment and qualified, trained personnel.
- ☐ Available product safety, first aid, physical property information.
  - Emergency telephone numbers.
  - Operations and emergency plans.

- Pollution and fire control.
- Equipment manuals.
- Product information sheets and brochures.

☐ Good engineering design.
- Equipment and operations should minimize contact of personnel with solvents.
- Protective guards on moving machinery.

☐ Observe operating personnel for safety awareness.
- Use of protective equipment.
- Performance of work.
- Proper use of tools. Tools in good condition.

☐ Periodic medical surveillance program.

☐ Training of terminal personnel and drivers.

FIRE AND ENVIRONMENTAL PROTECTION

☐ Check adherence to local fire codes.

☐ Fire protection plan. There should be a well thought out, written plan, and training drills, with specific assignments.

☐ Suitable fire fighting agents available; water, dry chemicals, foam. Appropriate foam for hydrocarbon and oxygenated solvents should be available.

☐ Fire extinguishers, hoses, hydrants accessible and in "ready" condition.
- Routine inspection program for fire equipment.
- Note condition of fire hoses, protective enclosures, racks.
- Check tags on equipment for date of last inspection.

☐ "No Smoking" signs prominently displayed and the rules enforced.

☐ Spill control procedures available.

HOUSEKEEPING AND MAINTENANCE

☐ Periodic maintenance inspection of all piping, tanks, valves, and equipment to avoid mechanical failures.

☐ Record keeping thorough, neat, and accurate.

☐ Work areas free of clutter, tripping hazards, weed growth, drainage.

☐ Condition of paint on tanks, and general condition of facilities.

☐ Check for spills around pumps and operations areas.

☐ Check disposition or storage of idle or obsolete equipment.

STORAGE OF SOLVENTS AT TERMINALS

SECURITY

- ☐ Condition of fences and gates.
- ☐ Screening of visitors and authorized personnel.
- ☐ Locks on valves affording access to products. This would include tanks, sample cocks, water draw-off valves, loading racks.
- ☐ Adequate off-hour vigilance.

Using a safety checklist of this type, one can learn to recognize the potential hazards of handling and storing solvents, and take appropriate precautions to minimize the hazards.

DETECTION AND PREVENTION OF EXPLOSIVE HAZARDS FOR SOLVENT VAPOURS

Udo Koss

Drägerwerk AG, P.O.Box 1339, 2400 Lübeck, W. Germany

Whenever solvents are transported, stored or used, explosive hazards can occur due to the inflammability of the solvent vapours in presence of a sufficient amount of oxygen and an ignition source. In order to avoid these hazards prevention measures have to be applied, which either

- substitute the inflammable solvent by a non-combustible solvent or
- restrict the concentration of inflammable vapours to a non-hazardous value or
- reduce the oxygen concentration of combustible solvent vapours or
- eliminate all possible kind of ignition sources.

In the following inerting and combustible vapour monitoring will be discussed in detail.

INERTING

Solvents are widely used in batch processes, e.g. in the pharmaceutical industries. The quantity of solvent to be used depends on the chemical process and can therefore not be altered in order to avoid explosion hazards. Vapours of these solvents may cause explosive hazards inside the apparatus, centrifuges, mills or other machinery. In these applications, the only way of eliminating the risk of explosive hazards is inerting, i.e. the replacement of environmental oxygen by a gas, which does not support combustion (nitrogen, carbon dioxide, etc.). As shown in table 1, a safe handling of solvents is possible, when the oxygen concentration does not exceed a certain limit, which is given by the type of solvent in use and by the kind of inert gas applied.

Table 1.

Maximum allowable O_2-concentration during inerting with N_2 or CO_2 in presence of different solvents

Combustible Substance	Maximum oxygen concentration (% O_2), when inerting with	
	N_2	CO_2
Benzene	11.2	13.9
Gasoline	11.8	14.5
Hexane	12.1	14.5
Pentane	11.6	14.4

Effectiveness of inerting has to be controlled in two directions:

a) the oxygen content of the solvent vapour / air / inert gas mixture must be kept below the maximum allowable oxygen concentration and
b) the amount of inert gas used for inerting should be as little as possible for economical reasons.

Both tasks mentioned above can be fulfilled by an instrument, which monitores the oxygen concentration in the container continuously and which controls the inert gas flow in relation to the oxygen concentration.
A multi-channel oxygen analyser with remote oxygen sensors has been developed for this application. The electrochemical oxygen sensors (polarografic measuring principle) have a low power consumption that the circiutry could be designed in an intrinsically safe manner. Thus they can be brought directly into the hazardous atmosphere inside the apparatus or equipment. Furthermore the intrinsically safe circuitry offers calibration by means of test gas at the sensor by a single person.
Figure 1. depicts a possible application of such an oxygen analyzer in an inerting process. The polarografic oxygen sensors are directly inserted into the vapour-containing equipment. The oxygen concentration is converted by the sensors into a DC-voltage. This DC-voltage is converted by the sensor electronics, which is directly connected to the sensor, into a frequency modulated signal for data transmission to the control console outside of the confined area. In the control console the oxygen concentration is indicated.

Switching contacts allow operation of different alarm measures, e.g. horns or rotating lights, and control - via solenoid valves - the flow of the inert gas.

Figure 1. Application of an oxygen monitor for inerting control

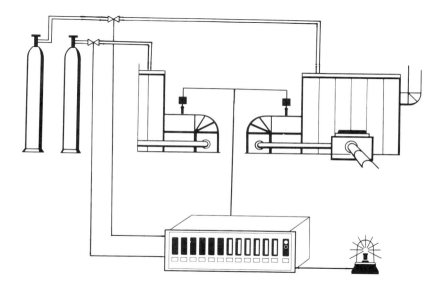

COMBUSTIBLE VAPOUR MONITORING

When solvents are handled at work-places, inerting is naturally not a measure to avoid explosive hazards. Ignition sources cannot be excluded completely in these applications, unless all equipment used would be built in safe design. For example, bulk-storage facilities are frequently used for solvents used in paint industry. Semi-automatic or automatic computer controlled storage handling equipment may cause damages to containers in these bulk-storage facilities. The risk of vapourizing those solvents would call for electrical equipment to be built in a flame proof design. Gas detection systems, which monitor the concentration of the combustible substance in the range below the lower explosive limit (LEL) to operate automatic alarm or even shut-down facilities, avoid such investment and allow the use of standard electric and electronic equipment respectively.

Gas detection equipment for combustible gases and vapours uses today the principle of catalytic combustion.

Figure 2. Combustible Vapour Monitoring - Measuring principle

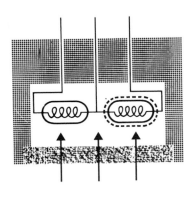

The gas / vapour-air mixture diffuses into the measuring chamber and is oxidized at the heated active pellistor. A second passive pellistor compensates for variations of ambient conditions (temperature, etc.). The difference, of the two pellistors connected in a Wheatstone bridge generates the output signal.

When monitoring the bulk-storage facilities it has to be kept in mind that several kinds of solvents have to be monitored. They may generate different signals at the sensor. Table 2. shows a example of inflammable solvents which may be stocked in paint industries. Practical measurements have shown a difference in sensitivity of up to factor 2 for some substances. For a convenient calibration of the monitor the different sensitivities and also the different flash points have to be observed. A compromise must be made between low sensitivity and low flash point for selection of a substance to calibrate the monitor giving a reliable response for all solvents stored.

Table 2.

Lower Explosive Limits and Flash Points of different Solvents used in Paint Industries

	LEL	flash point (°C)
Toluene	1,2	6
Ethanol	3,5	12
Ethylacetate	2,1	- 4
Methylethylketone	1,8	- 1
Xylol	1,0	30
Isopropanol	2,0	12
Acetone	2,5	- 20
Isobutanol	1,4	35

Some substances cause the risk of catalytic poisoning of the sensor, which results a loss of sensitivity. In order to find out drops in sensitivity caused by sensor poison, the measuring heads have to be checked with calibration gas in regular intervals.

For automatic calibration control a gas detection system was equipped with a calibration check system. It supplies the measuring heads via a small tube with calibration gas. Methane is used as calibration gas; solvent vapours are not suitable for these automatic calibration procedures because of possible condensation of the vapours in the tubes.

Methane can be used as calibration gas, when the correlation factor for the substance under investigation is known, which is determined experimentally.

During the calibration process, the alarm- and shut-down facilities of the gas detection equipment are inhibited. A calibration circuitry checks the individual measuring heads for sensitivity and - in presence of clean air - for zero drift. If deviations are detected, a fault alarm is triggered and the deviation of zero or sensitivity is indicated for each individual measuring head. With such a mircocomputer controlled device the calibration check is carried out and a reliable monitoring of explosive mixtures is possible.

Figure 3. Schematics of a multi-point calibration check for a gas detection system

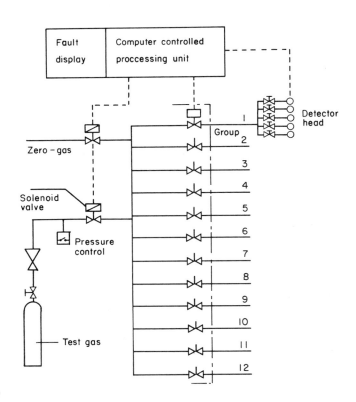

DETECTION OF EXPLOSIVE HAZARDS

SUMMARY

If a generated explosive gas / vapour-air mixture cannot be substituted by a non-combustible substance, explosive hazards can be prevented by inerting. Inerting can be controlled by an oxygen monitor.
      At work-places, generation of gas / vapour-air mixtures can be monitored by a gas detection instrument. Loss of sensitivity of the measuring heads can be detected early by an automatic calibration check system.

RECENT DEVELOPMENTS IN THE TOXICOLOGY OF SOLVENTS

Robert A. Scala

Exxon Corporation, Medicine and Environmental
Health Department, Research and Environmental
Health Division, P. O. Box 235,
East Millstone, New Jersey 08873, U.S.A.

INTRODUCTION

Mr. Chairman, ladies and gentlemen. Thank you for the opportunity to discuss with you some recent work involving the toxicology of solvents. This is a subject which has occupied a great deal of the attention of our group for the past six or seven years. It is also a subject which, to the classical toxicologist, presents a methodological problem--studies in humans are generating data of a sort which are not easily related to possible predictive or confirmatory work in laboratory animals. More on that later.

Sponsorship of this Symposium on the Safe Use of Solvents by an international body, IUPAC, is timely and appropriate. We find that the subject is clearly ripe for review with much active work in the past few years. We also find that the investigations are truly international in character. Important studies have emerged from Japan, the U.S., and Europe. There has even grown up a sort of international work group on the toxicology of solvents affording producers and users a forum for exchange of data on health effects research.

The remaining speakers in this morning's session will bring together interesting new work on two major classes of solvents (hydrocarbons and chlorinated compounds) and three classes of effects (reproductive hazard, carcinogenicity and the consequences of abuse). My intent will be to highlight some of the important recent work in the field.

This overview task seemed fairly straightforward. One need only review the literature and organize and summarize the major developments. For example, "The first task of toxicology is analysis. It must analyze the often very involved toxic pictures, discover the points of attack of

the poisons, clarify the mechanism of effect. The final and
most important goals of its work are the ascertaining of
dangers due to chemical materials, the prevention, recognition and treatment of chemical injuries." "Biological
experimentation is indispensable for investigation and
evaluation of the effects of solvents. Only animal experiments permit exact mathematical comparisons between individual materials and make possible exact knowledge of the
peculiarities of each material. By so doing, it gives us
important contributions and a basis for evaluation in
practice." Also, "In contrast with other industrial poisons
such as metals, e.g. lead and also arsenic, the conditions
for the accumulation of solvents in the body over a long
period of time are generally not favorable. They are for
the most part eliminated through the lungs more or less
quickly, depending on the volatility and solubility. However, there are solvents that are retained for several days
unchanged in the organism." "Many solvents, and among them
especially the most toxic ones, undergo a chemical change in
the organism. The resultant materials differ fundamentally,
both chemically and physically, from the solvents originally
absorbed. The new materials are more capable of reaction,
are of stronger polarity and of entirely different electrical behavior and their volatility, distribution, absorption
and solubility relations are different. Consequently, the
behavior of these changed products in the organs, cells and
body fluids is fundamentally changed." (Lehmann and Flury,
1943 (1)). The only problem with this summary of recent
work is that it was published in 1938 by Ferdinand Flury
from Wurzburg, and these quotations were from the 1943
translation of Lehmann and Flury's book on solvents.

One of the major contemporary concerns about hydrocarbon
solvent effects has been the problem of central and peripheral polyneuropathy. Like most of the audience, this
observer first became aware of the problem through the 1971
case reports by Herskowitz, Ishii and Schaumburg (2), which
also made reference to Japanese studies by Yamamura (1969)
(3). However, Browning (1965) (4) quotes a case from the
1933 literature of an individual who "developed such weakness of his arms and legs that he was unable to walk or
stand unsupported. There was some atrophy of the muscles
and severe polyneuritis was diagnosed". The individual had
been ingesting a hydrocarbon blend comparable to motor
spirit (gasoline) twice a day for 5-6 weeks in the belief
that it was a cure for venereal disease. Other investigators have cited cases presenting a picture of polyneuropathy
as early as 1856.

Browning also almost identified the relationship of

halogenated hydrocarbons and endogenous or exogenous catecholamines to ventricular fibrillation, resulting in sudden deaths with no apparent organic lesions in her description of cases from 1943-1951. This issue occupied many investigators a decade ago.

You, therefore, see before you a student of solvent toxicology who now believes that most, if not all, of what we will discuss today was anticipated by those careful early investigators who observed, recorded and speculated but who lacked the elegant analytical and biomedical instrumentation which characterizes current research.

This paper will be divided into two major parts--a discussion of newer methods of investigation in both animals and man and an overview of some of the more important effects, again using animal models and clinical reports. The literature cited is principally from 1980 and 1981.

Many of the studies to be reported involve complex designs and extensive use of subjective measurements. The amount of data generated is large and permits more than one interpretation. In this paper every effort will be made to report the investigators' findings and interpretations faithfully. On occasion, a personal view will be expressed principally to point to a limitation on the applicability of the data. At the outset it should be noted that there are widely held differences of opinion on the meaning of many of the studies in humans. These studies, furthermore, rarely provide any quantitative link to exposure levels of the solvents or other possible causative agents involved. Finally, faithful observance of TLV's, attention to accepted work practices and adequate personal hygiene should eliminate or control any health risk associated with solvent use. No unequivocal information to the contrary is at hand.

NEWER METHODS OF INVESTIGATION

In the past two years, five areas of methods research involving solvents have seen important progress. These include sensory irritation, dermal effects, nervous system studies in animals, biomonitoring and investigations in man.

The laboratory of Prof. Yves Alarie at the University of Pittsburgh (USA) has yielded a rich harvest of studies on a variety of industrial chemicals. One facet of their program has dealt with an animal model for predicting sensory irritation. Originally developed in connection with studies on potent chemical agents, a recent paper (Kane, et al., 1980 (5)) applied the method to 11 solvents and related the findings to the TLV for man. The method involves 10-minute exposures of four mice to vapors of the agent of interest.

Using a body plethysmograph, the respiratory rate is measured, and the RD_{50} determined. This is the concentration which produces a 50% reduction in respiratory rate (a reliable measure of irritancy). Earlier, Alarie had published a guideline showing the relationship between the RD_{50}, the expected human response and a proposed relationship to environmental standards (Table 1).

TABLE 1

RELATIONSHIPS OF RD_{50} CONCENTRATION VALUES TO INDUSTRIAL AND ENVIRONMENTAL STANDARDS AND PREDICTED EXPECTED RESPONSE IN HUMANS

CONCENTRATION	EXPECTED RESPONSE IN HUMANS	PROPOSED RELATIONSHIPS TO INDUSTRIAL OR ENVIRONMENTAL STANDARDS
10 RD_{50}	LETHAL OR EXTREMELY SEVERE INJURY TO THE RESPIRATORY TRACT	—
RD_{50}	INTOLERABLE SENSORY IRRITATION; RESPIRATORY TRACT INJURY MAY OCCUR WITH EXTENDED EXPOSURE	—
0.1 RD_{50}	DEFINITE BUT TOLERABLE SENSORY IRRITATION	HIGHEST ACCEPTABLE CONCENTRATION FOR TLV 0.2 RD_{50} - BASIS FOR STEL 0.3 RD_{50} - BASIS FOR EEL
0.01 RD_{50}	MINIMAL OR NO SENSORY IRRITATION	LOWEST CONCENTRATION NECESSARY FOR TLV
0.001 RD_{50}	"SAFE" LEVEL OF NO EFFECT	RECOMMENDED HIGHEST CONCENTRATION FOR AIR QUALITY STANDARD

FROM L.E. KANE, ET AL., AIHAJ 41:451, 1980

Seven of the materials were a homologous series of alcohols, and sensory irritation increased with an increase in carbon number. The authors then compared the RD_{50} values with the published TLV to see whether the published value was in the range predicted by the model. In 9 cases of 11, the TLV was in the range predicted. Assuming the validity of this approach, the TLV for methanol was overly conservative, perhaps reflecting a different basis for the TLV and ethyl acetate was not conservative enough, suggesting that, for some workers, the TLV could present an irritation problem. Stated differently, the test appears to be sensitive, but enough data are not at hand to judge specificity. The data are summarized in Table 2.

Another sensory irritation study was conducted for the American Petroleum Institute by L. Hastings, et al. (6) and seen as an undated draft report. The methods used incorporated both sensory response (as eyeblink, respiratory rate and eye-nose-throat irritation) and psychomotor performance (Michigan Eye-Hand Coordination Test, Choice Reaction Time and a visuomotor skill game) with young adult volunteers.

None of the subjects were cigarette smokers. Thirty-minute exposures to three hydrocarbon solvents of varying aromaticity at 1, 2, 3 and 4 times the TLV produced no effects on respiration or nose-throat irritation and some dose related effects on eyeblink and eye irritation. The psychomotor tests yielded large intersubject variability and, judged by performance ratios, showed no differences from baseline. The variability problems seemed to limit the utility of the studies and prevented the discerning of differences which might be present.

TABLE 2

RD_{50} VALUE, TLV VALUES AND RELATIONSHIPS TO RANGE PREDICTED BY MODEL

CHEMICAL	RD_{50} PPM	TLV PPM	TLV BETWEEN 0.1-0.01 RD_{50}?
METHANOL	42,000	200	NO
ETHANOL	27,000	1,000	YES
ISOPROPANOL	18,000	400	YES
n-PROPANOL	13,000	200	YES
n-BUTANOL	4,800	50	YES
ISO-PENTANOL	4,400	100	YES
n-PENTANOL	4,000	—	YES
ACETONE	78,000	1,000	YES
ACETALDEHYDE	4,900	100	YES
2-BUTOXYETHANOL	2,800	50	YES
ETHYL ACETATE	600	400	NO

FROM L.E. KANE, ET AL. AIHAJ 41:451, 1980. RD_{50} DATA HAVE BEEN ROUNDED.

Lupulescu and Birmingham (1976) (7) have applied the dramatic visualization capability of scanning electron microscopy (SEM) to the study of the utility of a protective agent on minimizing skin effects from a lipid solvent. Both acetone and kerosine were applied undiluted to the skin of human volunteers for up to 90 minutes. Some of the sites were also treated with a water/glycerin/cellulose-methasol gum preparation. Punch biopsy specimens were prepared for light, electron and scanning electron microscopy. The results for kerosine are shown in Figures 1, 2 and 3. Clinically, kerosine produced intense hyperemia and the SEM shows edema, cellular disorganization, lacunar formation and damage to corneocytes. The hair shaft was not affected. Use of the protective gel resulted in a generally normal appearance. This work provides a means to guide the development of skin protective agents and measure their effectiveness against various classes of solvents.

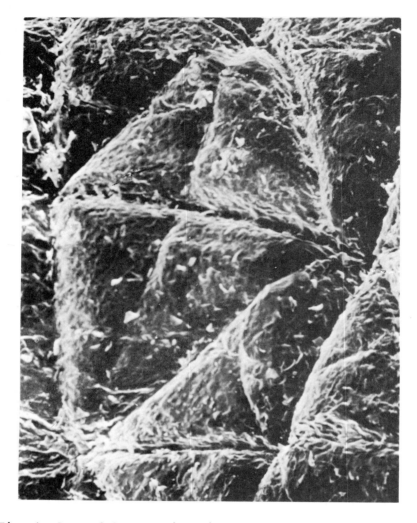

Fig. 1. Control human epidermis - SEM 270X

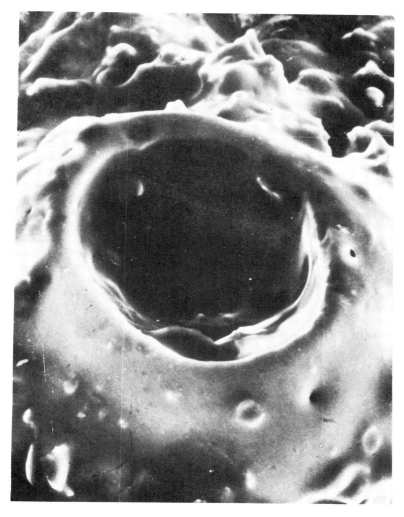

Fig. 2. Human epidermis after kerosine - SEM 360X

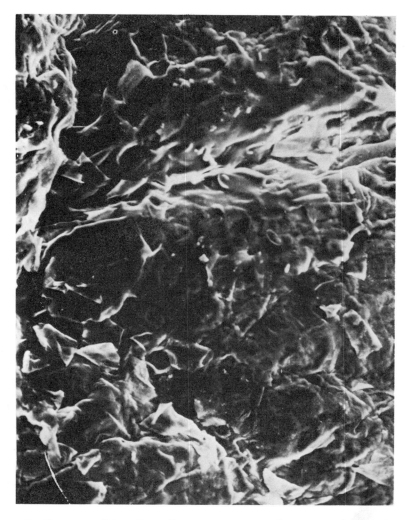

Fig. 3. Human epidermis after acetone and protective gel - SEM 360X

In the workplace, increasing attention has been paid to personal monitoring as contrasted with area monitoring to measure solvent exposure. Associated with this is the concept of the worker as his own monitoring device. To do so requires knowledge of uptake and elimination and how these are influenced by respiratory rate, exercise and amount of adipose tissue. Two recent papers have dealt with this issue (Gompertz, 1980 (8) and Brugnone, et al., 1980 (9)). Gompertz notes that genetically and environmentally determined pharmacokinetic factors control rates of clearance and metabolism. Knowing the time course of metabolite production and excretion will help to set the sampling strategy. If the objective is to obtain an alveolar air sample (which is in equilibrium with venous blood) assumed to be representative of time-weighted average exposure, then the time of sampling should have a known relationship to maximum and equilibrium blood levels. Disregard of such kinetic information will, Gompertz asserts, provide unrepresentative samples. Brugnone, et al. have published data on 10 solvents in which alveolar air samples were compared with breathing zone samples. The solvents studied were toluene, styrene, methyl ethyl ketone, acetone, dimethylformamide, cyclohexane, n-hexane, methyl cyclopentane, 2-methyl pentane and 3-methyl pentane. The ratio of alveolar air level to workplace concentration tends to be <0.5 for materials with high solubility in blood and >0.5 for materials with low solubility (and low retention) (Table 3). This method does not seem to have been applied by others.

TABLE 3

ALVEOLAR AIR SAMPLES FOR BIOMONITORING

C_a - CONCENTRATION IN ALVEOLAR AIR
C_i - CONCENTRATION IN WORKPLACE AIR
($C_i C_a$) CORRELATES WITH C_i:
  WHERE SOLUBILITY IN BLOOD IS HIGH $\frac{C_a}{C_i} < 0.5$
  WHERE SOLUBILITY IN BLOOD IS LOW $\frac{C_a}{C_i} > 0.5$

FROM BRUGNONE, ET AL. (1980)

The two centers from which most reports have issued on nervous system effects in humans from solvents are the Karolinska Institute in Stockholm and the Institute of Occupational Health in Helsinki. Gottfries, et al. (1980) (10), in an unpublished symposium paper, outlined in some detail the methodology which they and their colleagues in Stockholm (Department of Occupational Health and Karolinska Institute) have used to study the effects of solvents on the nervous system. They have found both central (CNS) and peripheral (PNS) system effects. The central effects included mental disorders, impaired physiological function, nerve cell damage and effects on cranial nerves. The scope of the examination (Table 4) involved psychiatric, neurological, neurophysiological and neuroradiological techniques. This group has shown considerable interest in neurasthenia as the dominant nervous system effect. They note that in the period 1966-1975 there have been 200 administrative determinations producing early retirement in Finland because of "psycho-organic syndromes" associated with solvent exposure. From their perspective, Gottfries, et al. feel CNS and PNS effects can occur following long-term exposure to solvents even at levels less than the official occupational exposure limits. Unfortunately, like most of this literature, their report contains no details of exposure levels to the various hydrocarbon, oxygenated and halogenated solvents associated with the reported nervous system changes. This reviewer is not inclined to believe that a disease or diagnosis has been established as much as a symptom collection, a constellation of findings or perhaps even an easy route to retirement.

TABLE 4

PROCEDURES FOR THE STUDY OF NERVOUS SYSTEM EFFECTS FROM SOLVENTS

PSYCHIATRIC
    QUESTIONNAIRES
    INTERVIEWS
    HEALTH EXAMINATION
    RATING SCALES
    PSYCHOLOGICAL TESTS
NEUROLOGICAL
NEUROPHYSIOLOGICAL
NEURORADIOLOGICAL

FROM GOTTFRIES, ET AL. (1980)

ADVERSE EFFECTS FROM SOLVENT EXPOSURE

The traditional view of solvent toxicity is stated simply. Topically, solvents are defatting agents on the skin, leading to drying, cracking and dermatitis. Solvents by the inhalation route are central nervous system depressants, acting somewhat like narcotic or anesthetic agents. The effects are not permanent unless there is the complicating issue of prolonged anoxia. Even in the era of Lehmann and Flury it was known that this was not solely the case but much of what has been done with respect to the safety of solvents in industrial use seems keyed to those tenets. The current experimental or clinical literature will reveal a range of additional effects, both organ specific and biochemical, associated with solvent exposure. The task before us is to relate these findings to compositional features and actual occupational exposure levels. In judging safety in the workplace, the importance of the latter is often not recognized nor reported. This deficiency characterizes many of the papers cited.

At the cellular level, it is probably no surprise that certain halogenated hydrocarbon solvents are activated by the cytochrome P-450 system in rat liver cells and will subsequently alkylate both lipids and proteins. In addition, the metabolites of carbon tetrachloride and trichloroethylene will also alkylate DNA and RNA (Cunningham, et al., 1981 (11)). Further work by these investigators reflected the stimulating action of phenobarbital pretreatment and the impact on binding following treatment with agents which reduce glutathione levels. Perhaps more surprising was the report of Kramer, et al. (1974) (12) that liver microsomal enzymes are fully capable of (ω-1) oxidation of hexane or heptane. Indeed, n-hexane is said to work as well as phenobarbital in inducing this reaction. Although the predominant site is the (ω-1) or C_2 position, oxidation by liver microsomal enzymes will also occur at C_1 and C_3. This was an important step in the chain of evidence (Scala, 1976 (13)) suggesting a common pathway in the formation of neurotoxic species from both n-hexane and methyl n-butyl ketone.

Other hydrocarbon solvents besides hexane also stimulate cytochrome P-450 activity. In an unpublished symposium paper, Gustafsson (1980) (14) exposed male rats to levels of xylene, methyl ethyl ketone and toluene (and other solvents) above the TLV. By use of biphenyl and benzo(a)pyrene as substrates, he showed that xylene and toluene were phenobarbital-like inducers of cytochrome P-450. MEK was without effect in this study.

Mention was made earlier of the central and peripheral polyneuropathy associated with exposure to n-hexane. In a brief teaching monograph published in 1979, Schaumburg and Spencer (15) summarized current knowledge of animal models of human nervous system disease in which the effects are elicited by administration of neurotoxic agents. In the section devoted to central-peripheral distal axonopathy, these authors review the findings in experimental animal intoxications, the clinical picture in humans occupationally exposed to these agents at high concentrations and the close correlation between the two. Since linear hexacarbon solvents are among the chemicals producing this effect in laboratory animals and man, a brief mention is in order. In laboratory animals, neurotoxic agents of this class produce a gradual onset of symmetrical hindlimb weakness and unsteadiness which, with continued administration, eventually affects the forelimbs as well. The doses are usually high and the route of exposure usually oral or parenteral for convenience. Severely intoxicated animals do not fully recover after removal of the active agent, showing residual spasticity and ataxia. The neuropathological picture shows multifocal degeneration in the distal portion of long and large diameter axons. With time and continued administration of the agent, these axonal changes proceed proximally toward the nerve cell body in a multifocal fashion, not clearly "dying back" as originally thought. The neurotoxic agent also produces distal fiber degeneration in the CNS with gracile, cortico-spinal and spino-cerebellar tracts the most susceptible. These CNS and PNS structural changes considerably predate the onset of neurological findings in experimental animals. Schaumburg and Spencer speculate that when humans first show signs of neurological impairment, there is already an advanced state of degeneration of some PNS and CNS axons. Much of what is seen in human cases can be correlated with the findings in experimental pathology.

TABLE 5

CORRELATION OF HUMAN CLINICAL FINDINGS AND EXPERIMENTAL PATHOLOGY - HEXACARBON NEUROPATHY

GRADUAL, SLOW ONSET - LOW LEVEL, STEADY EXPOSURE AFFECTS DISTAL POSITION OF SELECTED FIBERS—PATIENT REMAINS FUNCTIONAL.
INITIAL FINDINGS IN LOWER EXTREMITIES - LONG AND LARGE AXONS AFFECTED EARLY. SCIATIC NERVE ESPECIALLY VULNERABLE.
STOCKING-GLOVE LOSS - DEGENERATION OF DISTAL MOTOR AND SENSORY AXONS PROCEEDS TOWARD CELL BODY—CLINICAL SIGNS INITIALLY IN HANDS AND FEET.
EARLY LOSS OF ACHILLES REFLEX - FIBERS TO CALF MUSCLES ARE VERY LARGE DIAMETER AND ARE AMONG FIRST TO BE AFFECTED.
MOTOR NERVE CONDITION - ONLY SLIGHTLY DECREASED AS MANY MOTOR FIBERS REMAIN INTACT. NCV ALMOST NORMAL DESPITE CLINICAL SIGNS.
SLOW RECOVERY - AXONAL REGENERATION IS ABOUT 1 MM/D (UNLIKE REMYELINATION). RECOVERY MAY TAKE MONTHS OR YEARS.
SIGNS IN CNS AFTER RECOVERY - CNS DEGENERATION MAY BE MASKED BY EARLY PNS CHANGES AND BE EVIDENT ONLY LATER.

FROM SCHAUMBERG, H.H. AND P.S. SPENCER, NEUROTOXICOL. 1:209-220 (1979)

This is outlined in Table 5. The pathogenesis of this disorder is still not fully understood. Certainly more seems to be involved than a direct effect of the neurotoxin on the nerve cell. The pioneering study of the morphological events by Spencer and co-workers has helped direct the ongoing research on the biochemical and physiological events which produce the axonal degeneration.

Although a later speaker in this symposium will discuss

hydrocarbon solvent toxicity and, in particular, n-hexane neuropathy in more detail, one further note might be made. With the reports issuing on hexacarbon neuropathy, considerable interest was expressed on how general this effect might be. Did, in fact, the potential for neurotoxic effects extend beyond n-hexane to other six-carbon isomers or to other chain lengths? A recent study by Egan, et al. (1980) (16) involved exposure of rats 22 hours per day for up to six months to 100 ppm methyl n-butyl ketone (MNBK), 500 ppm methyl ethyl ketone (MEK) and 500 ppm of a commercial hexane mixture which contained no n-hexane (Table 6). Examinations included detailed neuropathology studies. Animals exposed to MNBK showed typical hexacarbon-induced neuropathy in CNS and PNS. MEK and the n-hexane-free hexane blend were free of neurotoxic properties.

TABLE 6

COMPOSITION OF n-HEXANE- "FREE" HEXANE

ISOMER	WEIGHT %	NOMINAL CONCENTRATION IN CHAMBER, MG/M^3
2,3-DIMETHYL BUTANE	3.4	60.0
2-METHYL PENTANE	35.3	618.0
3-METHYL PENTANE	30.0	525.0
METHYL CYCLOPENTANE	24.6	431.0
n-HEXANE	0.3	5.3
CYCLOHEXANE	6.2	109.0

FROM EGAN, ET AL. NEUROTOXICOL. 1:515-524 (1980)

Independent of these studies on the neurotoxic potential of hydrocarbon and oxygenated solvents, another series of experiments was undertaken to provide basic data on the inhalation toxicology of a series of widely used hydrocarbon solvents and to provide guidance in the establishment of occupational exposure limits. Some of this work was undertaken by a major trade association on behalf of its members. The results of this work have been published in a series of 17 papers by Carpenter, Weil and co-workers (1975-1978) (17). Other work was undertaken on a joint basis by interested companies and will be reported later in this meeting. Individual companies also carried out animal experiments on solvents of particular interest to them. In one series, rats were exposed 6 hours per day, 5 days per week for 13 weeks to four hydrocarbon solvents of varying composition (Table 7) at concentrations ranging from 100 to 1200 ppm. The animals showed evidence of renal tubular degeneration confined to the corticomedullary region. Also seen were focal dilation and granular eosinophilic debris.

There was no effect on the glomerulus. The effect was
confined to male rats (Lione, 1979, 1981) (18).

TABLE 7
**COMPOSITION OF SOLVENTS
PRODUCING NEPHROPATHY**

SOLVENT	BOILING RANGE	COMPOSITION, WT. %		
		AROMATIC	PARAFFINS	NAPHTHENES
A	98-105	0.02	99.9	0.1
B	160-175	0.03	99.9	0.1
C	148-198	0	52	48
D	148-196	21	55	24

The earlier studies of Carpenter, et al. (1975-1978) (17)
were reviewed and kidney sections from those studies were
reexamined. These studies were also of 13 weeks duration
and included both rats and dogs. Of the 16 additional
materials at hand, 6 showed the same kidney lesion in male
rats, 6 showed no evidence of damage and 4 others were
tested at such low levels because of their inherent low
volatility that the negative results are difficult to
interpret. Also available for review were sections from a
trade association sponsored study of unleaded motor gasoline
which contains many of the components found in hydrocarbon
solvents. Rats and mice were exposed to vaporized, full
boiling range motor fuel (including additives) at levels of
50, 275 and 1500 ppm. Data were available through 18 months
of a scheduled 24-month experiment. The findings (Lione,
1981) (18) are summarized below.

- Regenerative epithelium in the renal tubule of male
  rats was seen as early as three months in all
  treatment groups.

- Renal tubular dilation with intratubular protein was
  seen as early as three months in male rats at the
  two higher exposure levels.

- Papillary mineralization was seen in the kidneys of
  male rats at the 275 and 1500 ppm exposure levels at
  12 and 18 months.

Subsequent to these reports, there was a further preliminary
report following the termination of this motor fuel study
(Weaver, 1981 (19)). Male rats from this study showed an
apparent increase in kidney tumors at all levels tested.

The findings with the solvents are shown in Table 8 and with motor fuel in Table 9. A large number of questions remain unanswered with respect to the nephropathy noted in rats.

TABLE 8

KIDNEY FINDINGS - MALE RATS - SOLVENTS

SOLVENT	BOILING RANGE °C	COMPOSITION			CONC. (ppm, max	RENAL TUBULAR DAMAGE
		Ar*	P*	N*		
VM&P	118-152	12	55	33	1,200	+
STODDARD	152-193	14	48	38	330	+
60 S	128-158	29	50	21	1,200	+
HI NAPH	155-182	1	29	70	1,000	+
NAPH-AR	150-200	38	25	37	790	+
70 S	155-210	16	58	26	410	+
N-NONANE	150	—	100	—	1,600	−
50 TH	98-105	66	33	1	600	−
RUBBER SOLV	66-112	5	41	54	2,000	−
TOL. CONC.	95-111	39	46	15	985	−
80 TH	97-142	10	71	19	390	−
XYLENES	137-141	100	—	—	810	−
140° FL ALIPH	183-206	61	3	36	37	?
KEROSINE	207-271	55	4	41	14	?
40 TH	186-231	35	32	33	36	?
HI AR	183-206	100	—	—	66	?

* Ar = AROMATICS   P = PARAFFINS   N = NAPHTHENES

FROM LIONE, 1981

TABLE 9

KIDNEY FINDINGS - MALE RATS - MOTOR FUEL

EFFECT	EXPOSURE LEVEL, NOMINAL PPM			
	0	50	275	1500
REGENERATIVE EPITHELIUM (MEAN GRADE)				
3 MOS	0	0.4	1.2	1.7
6 MOS	0.2	0.6	1.6	2.5
12 MOS.	0.1	1.1	0.8	1.0
18 MOS.	2.4	2.3	3.4	3.4
PAPILLARY MINERALIZATION ANIMALS				
3 MOS	0	0	0	0
6 MOS.	0	0	2/12	8/11
18 MOS.	0	0	1/11	10/13
RENAL TUMORS (NUMBER)				
24 MOS.	0	2	6	5

DATA FROM LIONE, 1981 AND WEAVER, 1981

Among them are the structural requirement (activity appears to be associated with paraffinic and naphthenic materials boiling above about 100°C), whether comparable effects occur in man, strong species and sex specificity, reversibility and relationship of pre-existing kidney lesions to the later development of mineralization and/or renal tumors.

In attempting to gain an understanding of the relevance of these findings in rats to potential hazard for man, we find that the literature has not provided much guidance. Looking just at renal tubular damage, there seems to be no body of data suggesting a similar effect in man apart from renal tubular acidosis associated with acute episodes of solvent abuse. One may speculate, however, that the glomerulus in man may be the more sensitive locus for possible solvent effects. There are case reports of chronic glomerulonephritis in man associated with chronic solvent exposure. van der Laan (1980) (20) failed to find an association but did include in his paper mention of three earlier studies in which there was an association reported between the disease and solvent exposure. These data are shown in Table 10.

TABLE 10

CASE-CONTROL STUDIES — CHRONIC GLOMERULONEPHRITIS AND ORGANIC SOLVENT EXPOSURE[1]

INVESTIGATORS	NUMBER OF CASES	NUMBER OF CONTROLS	% CONFIRMED (BIOPSY)	CLINICAL STAGE	RELATIVE RISK
ZIMMERMAN, ET AL., 1975	37	63	57	ADVANCED	>4
LAGRVE, ET AL., 1977	247	124	76	NEW AND ADVANCED	5.3
ROVNSKOV, ET AL., 1979	50	100	100	NEW	3.9
VAN DER LAAN, 1980	50	50	100	NEW	1.1

FROM VAN DER LAAN, INT. ARCH. OCCUP. ENVIRON. HEALTH 47:1-8, 1980

[1] SOLVENTS NOT IDENTIFIED; EXPOSURE DATA NOT GIVEN

The actual solvents with which the individuals worked were not stated nor was there any quantitative exposure information. The suspicion is that exposures were high. The workers were grouped into heavy, moderate and light exposure categories based on the task. For example, indoor house painting was in the heavy exposure category; outdoor painting was a light exposure. There was no relationship between the disease and specific solvent types nor was any histological or clinical type of chronic glomerulonephritis represented to the exclusion of others.

Olsson and Brandt (1980) (21) reported on a possible association between occupational exposure to organic solvents and Hodgkin's disease. This is not the first report of such an association for solvents or other chemicals. The authors do suggest that solvents (halogenated, oxygenated and hydrocarbon) are an especially hazardous class. In their case-referent study, 25 men (age 20-65) with Hodgkin's disease were matched with 50 sex, age and resident controls. Each individual was interviewed to determine occupational exposure to solvents and other chemical agents. Exposure was defined (everyday contact for at least one year in the ten years prior to interview) but not measured. Only solvent exposures were considered. No mention was made of other chemicals. Smoking and drinking habits showed no difference. A relative risk of 6.6 was calculated based on 12/25 Hodgkin's patients with occupational exposure to solvents and only 6/50 referents (Table 11). Of the 12 Hodgkin's patients, 9 were exposed to aromatic hydrocarbons. The list of solvents included xylene, toluene, hexane, tetrahydronaphthalene, white spirit, cyclohexane, styrene, trichloroethylene, glycol ether, acetates, ethanol, methanol and hydroquinone. In view of the small sample size and lack of exposure data, this study must be considered no more than suggestive.

TABLE 11

OCCUPATIONAL EXPOSURE TO ORGANIC SOLVENTS AND HODGKIN'S DISEASE IN MEN
RESULTS OF A CASE-REFERENT STUDY

	REFERENT EXPOSURE		
	BOTH	ONE	NEITHER
PATIENT EXPOSED	—	3	9
PATIENT NOT EXPOSED	—	3	10

FROM H. OLSSON AND L. BRANDT. SCAND. J. WORK ENVIRON. HEALTH 6:302-305 (1980)

Much of the preceeding discussion of actual or suggested adverse effects in humans has only set the stage for what is judged to be the most controversial and troublesome group of adverse findings--nervous system changes in man not clearly diagnosable as organic disease. This collection of reports deals with neurological, neurophysiological, psychomotor and psychiatric changes in workers exposed to a variety of solvents at or below the recommended occupational exposure limits. The subject is controversial in that the association is primarily circumstantial and generally not confirmed by controlled animal studies nor are the diagnoses unique. There is, in fact, a question whether a disease or merely a "symptom collection" is involved. Juntunen, et al. (1980) (22) acknowledge this, noting that "occupational disease caused by long-term exposure to solvents is always a probability diagnosis, based much on the exclusion of other diseases causing similar symptoms."

In response to a published query on how to diagnose "solvent poisoning", Seppalainen and Lindstrom (1981) (23) noted that (as with any toxic encephalopathy) no complete clinical description exists. The examining physician should consider the neurophysiological and psychological findings; the symptoms relating to malfunction of the peripheral or central nervous system and ascertain clearly that occupational exposure to solvents did occur. Though not mentioned specifically, it is equally important to consider the temporal relationship of exposure and findings and exposure to other agents such as heavy metals, pigments, resins, reactive chemicals and the like. One need also question whether these "effects" constitute "disease" or simply variances within a normal range.

The troublesome element arises in the implication that currently accepted official occupational exposure limits apparently do not protect workers from these nervous system effects. Such studies also edge close to the Eastern European standards of "effect".

By way of contrast, the evidence seems quite incontrovertible that abusive exposure to solvents as seen in glue sniffing or "huffing" will produce unmistakable nervous system damage. Among the most recent papers in this extensive literature is a report from Glascow (King, et al., 1981 (24)). King suggests that glue sniffing be entertained as a diagnosis when a child experiences unexplained coma, convulsions, ataxia, and behavior disturbances. The novel feature of this work is a proposed correlation between toluene levels in blood and the clinical picture (Table 12).

TABLE 12
PROPOSED RELATIONSHIP BETWEEN BLOOD LEVELS OF TOLUENE AND CLINICAL SIGNS

BLOOD LEVEL (MG/G)	CLINICAL FEATURES
0	
0.5	INDUSTRIAL EXPOSURE
1.0	NOTED ON BREATH
5.0	
20	SEVERE SIGNS
60	PROBABLY FATAL

FROM M.D. KING, ET. AL. BRIT. MED. J., 1981

King, et al. reason that toluene is absorbed from the lungs and bound to lipoprotein. Metabolism occurs to the extent of 70-80% of the absorbed dose by hepatic microsomal enzymes. With its high affinity for lipid tissues, the blood concentration is biphasic demonstrating an initial peak, then a trough reflecting CNS binding, and a subsequent peak as the toluene is slowly released into the blood. In fatal cases, the CNS concentration exceeds that of blood. A discussion of glue sniffing is not complete without some mention of the work of Altenkirch, et al. (1977) (25) who clearly related composition, clinical features and nervous system histopathology. Their work also dealt with the subject of solvent interaction in producing nervous system disease. This point will be noted again later.

These introductory observations lead to a series of case reports primarily from Scandinavia on nervous system effects of solvents. The consistent features in these publications are the constellation of occupational exposure to solvents (usually unspecified and ordinarily uncontrolled), non-specific polyneuropathy symptoms, some neurophysiological changes and evidence of psychiatric and psychomotor alterations--the latter two being the most persuasive elements.

Fazio and Ballistini (1979) (26) conducted medical and neurological examinations on 670 wokers in 24 small plants in Italy. Sixty subjects had polyneuropathy. EEG abnormalities were non-specific, appeared later than peripheral symptoms and reversed more slowly. Of greatest interest was the presence of anxiety-depressive symptoms including sleep disorders, free floating anxiety, depressed mood, autism, decreased libido and somatic complaints (Table 13). These were discerned by questionnaire and correlated with conditions of work including ventilation in the workplace and estimates of exposure level. However, no information was given on specific solvents or airborne concentrations. The levels in these small plants must have been very high based on work of others.

TABLE 13

ANXIETY-DEPRESSIVE SYMPTOMS REPORTED IN SOLVENT EXPOSED WORKERS

SLEEP DISORDERS	**FREE FLOATING ANXIETY**
DEPRESSED MOOD	**AUTISM**
DECREASED LIBIDO	**SOMATIC COMPLAINTS**

FROM C. FAZIO AND N. BALLISTINI, 1971

The impact of actual workplace conditions was neatly illustrated in a report of reaction time changes in solvent-exposed workers (Olson, et al., 1981 (27)). There was an opportunity to make measurements of CNS function (reaction time) in a plant where work practices resulted in exposures to methyl ethyl ketone (MEK) regularly up to three times the TLV and during a weekly cleaning operation, up to 8-10 times the TLV. After certain equipment and procedural changes were made, the MEK level was about 1/5 the TLV except during some operations when it was about 0.7 TLV. Reaction time in the group of workers (19 out of 42 originally studied) improved with each improvement in work practices and equipment. Olson, et al. believe the observed CNS effect of delayed reaction time from solvent exposure was reversed as exposure levels decreased.

The group at the Karolinska Institute in Stockholm have emphasized psychiatric interviews, and rating scales are considered preferable to questionnaires in the quantitative assessment of mental symptoms (Struwe, et al., 1980 (28)). They believe that ratings by trained interviewers are as useful as many neurophysiological and psychometrical methods to detect early neurotoxicity. Furthermore, slight changes in "mental comfort" and personality often are reported to be the first signs of progressive brain damage. The principal rating tool utilized was the Comprehensive Psychopathological Rating Scale and was employed in an in-depth evaluation of eighty painters with long-term, low-level solvent exposure (Elofsson, et. al., 1980 (29)).

TABLE 14

NERVOUS SYSTEM EFFECTS IN PAINTERS

CATEGORY	TESTS	FINDINGS
PSYCHIATRIC	44 ITEM SCALE	SLIGHT NEURASTHENIC SYNDROME
PSYCHOLOGICAL PERFORMANCE	18 TESTS	CHANGES IN REACTION TIME, MANUAL DEXTERITY, SHORT TERM MEMORY, PERCEPTUAL SPEED
NEUROLOGICAL AND NEUROPHYSIOLOGICAL	VER EEG EMNG VIBRATION SENSE	CHANGES IN LONG SENSORY FIBERS
OPHTHALMIC	LENS CONDITION	"CHANGES NOTED"

FROM S.A. ELOFSSON, ET AL. (1980)

These workers were compared to two matched, unexposed groups. The solvents involved a variety of hydrocarbons, oxygenated compounds and halogenated hydrocarbons. These workers were also exposed to pigments including chromium and lead, dust from grinding and degreasing agents. Exposure was assessed by interview and measurements. The multifaceted approach included psychiatric interviews, psychometric tests, neurological, neurophysiological and ophthalmologic examinations and computerized brain tomography. The time-weighted average exposures for the solvents (including combination exposures assuming simple additive action) were on the order of 1/5 the Swedish TLV and ranged from 0.01 TLV to 0.57 TLV (for halogenated hydrocarbons). The findings are noted in Table 14. The degree of exposure and extent of effect did not seem to be correlated. The subjects were all fit for work. The responses were distributed within the clinically normal range. This is a continuing puzzle to classically trained experimentalists who rely heavily on dose-response relationships in attributing causality. The persistent non-specificity of the psychopathologic symptoms is the central conclusion of an East German study of solvent exposed workers (Seeber, et al., 1979 (30)). It does seem, however, that even if the symptom picture is non-specific, the reported effects are beginning to aggregate into system classes. Unmentioned in most studies are the confounding variables of alcohol and drug usage. In addition, hobby and other off-the-job activities can yield significant exposure to chemical agents not normally appreciated unless specifically sought by the investigator. This is particularly evident in the work emerging from the group at the Institute of Occupational Health at Helsinki. Seppalainen, et al. (1980) (31) have found neuropsychological and psychological tests to be helpful in the evaluation of the hazards of solvent exposure. They studied 107 patients with a diagnosis of "solvent poisoning" following long occupational exposure. The patients were selected for study on the basis of existing diagnosis of occupational disease associated with exposure to organic solvents. The solvents were principally halogenated hydrocarbons but also included aliphatic and aromatic hydrocarbons and "paint solvents". The exposure levels were graded as low (generally low concentration or occasional exposure; repeated high-peak levels improbable), intermediate and high (close to the Finnish TLV but seldom exceeding it on a time-weighted average basis). The assignment of exposure levels was based on information supplied by the worker, the employer or "at times" actual measurements. Affected workers were assigned to all three grades, more to the "intermediate" and "high" than "low". They have found, in general, that occupational exposure to solvents and mixtures produces a constellation of effects consisting of polyneuropathy, organic psychosyndrome, EEG abnormalities and adverse effects on intellectual or memory functions or psychomotor performance. The profile of this patient group and the principal findings are shown in Table 15. Seppalainen, et al. found some relation between long exposure and poor performance in psychomotor tests. They discussed the issue of exposure length and degree and suggested that level may be more important than duration in determining the extent of damage, a conclusion not shared by other authors. All this work suffers from a lack of quantitative exposure information.

TABLE 15

NERVOUS SYSTEM EFFECTS ON SOLVENT POISONING

POPULATION	48 MALES	59 FEMALES
AGE, YEARS	35.8 ± 11.4	42 ± 10.3
EXPOSURE, YEARS	9.6 ± 8.6	7.6 ± 7.5

TEST RESULTS
HIGH INCIDENCE OF ABNORMAL EEG AND SLOWED NERVE CV
LOWER SCORES IN SUBTESTS OF ADULT INTELLIGENCE TESTS

FROM SEPPALAINEN, ET AL. AMER. J. IND. MED. 1:31-42 (1980)

A companion paper to this one (Lindstrom, 1980 (32)) focussed on psychological performance. The solvent exposed group was characterized by a decline in visuomotor performance and "decreased freedom from distractability". Here, duration of exposure seemed more important than level. Intellectual performance did not decline.

Husman (1980) (33) and Husman and Karli (1980) (34) examined a group of 102 car painters with low-level exposure to organic solvents. In this case control study, a comparable group of 102 referents was also used, and the study was conducted by using a symptom questionnaire. The painters were employed in auto repair shops in Helsinki and were exposed to paint binders (resins), dyes, pigments, fillers, catalysts and aromatic and aliphatic hydrocarbons, alcohols, esters and ketones. The organic solvent levels were measured in about 20% of the shops, and the concentration was about 1/3 the TLV. Fatigue and disturbances in memory and vigilance occurred more frequently among the painters than the referent group. A full clinical neurological examination was made of the two groups; the main deficit noted was sensory and involved vibration sense, light touch and pricking pain. No n-hexane or MNBK was used in these shops, yet the findings suggest peripheral neuropathy in long sensory fibers.

Juntunen, et al. (1980) (22) also addressed the question of the non-specificity of the symptoms in solvent affected patients. They emphasize the role of individual constitutional factors, such as metabolic capacity and physiological functions, in determining the precise clinical picture of disease. They studied, in-depth, 37 patients (out of 464 persons diagnosed with occupational disease due to organic solvents between 1964 and 1976). This sub-group had relatively severe symptoms and were submitted for pneumoencephalographic examination. There was little information on actual workplace concentrations, but the exposures were to a variety of solvents including carbon tetrachloride, trichloroethylene, carbon disulfide and others. They found EEG changes ("slight diffuse slow wave"), slight ENMG changes suggesting peripheral neuropathy in 23/28 so examined and psychological alterations in all, including personality changes (94%), psychomotor disturbances (80%), as well as changes in memory and learning tests (69%) and intelligence tests (57%). No clear-cut exposure effect relationship was seen.

Juntunen, et al. also discuss the thorny question of diagnosis. They, too, acknowledge the vague, nonspecific nature of the early symptoms. This makes the study of organic solvent intoxication by means of epidemiologic methods difficult. The onset of neurological symptoms cannot be accurately measured, hence the latency is not ascertainable. The very nonspecificity requires extreme skill in determining what to include and what to exclude. The diagnostic criteria used in Finland are (Table 16): verified exposure to neurotoxic agent; a clinical picture of organic nervous system damage; exclusion of other organic diseases and exclusion of primary psychiatric diseases.

TABLE 16

DIAGNOSIS OF ORGANIC SOLVENT POISONING

- **VERIFIED EXPOSURE TO NEUROTOXIC AGENTS**
- **CLINICAL PICTURE OF ORGANIC CNS AND/OR PNS DAMAGE**
    — **SUBJECTIVE SYMPTOMS**
    — **CLINICAL NEUROLOGICAL TESTS**
    — **EEG/ENMG**
    — **PSYCHOLOGICAL TESTS**
- **OTHER ORGANIC DISEASES REASONABLY WELL EXCLUDED**
- **PRIMARY PSYCHIATRIC DISEASES REASONABLY WELL EXCLUDED**

FROM JUNTUNEN, ET AL. (1980)

Among the most efficient criteria in making the diagnosis of organic solvent poisoning are peripheral sensory neuropathy, psycho-organic alteration, visuomotor function and psychomotor performance. Individual differences in susceptibility to chemical agents are seen, and some patients experienced a deterioration in function despite cessation of exposure and others recovered fully.

One additional item remains in this review of recent work on solvent induced disorders. That relates to the question of possible interaction or even synergy between components present in solvent blends. Much of this attention has focussed on the role of methyl ethyl ketone (MEK). Several papers have reported on the apparent enhancement of the neurotoxicity of methyl n-butyl ketone (MNBK) in experimental animals by the simultaneous exposure to MEK. Most striking is the decrease in latency for clinical or histopathological evidence of polyneuropathy. The work of Altenkirch, et al. (1977) (25) is highly supportive of an enhancing effect of MEK on the neurotoxicity of n-hexane. The basis for their study was a group of eighteen juveniles who were sniffers of glue thinner and who developed a toxic polyneuropathy. All the usual diagnostic criteria for the neuropathy were met--symmetrical, progressive, ascending, mainly motor polyneuropathy with muscle atrophy. Biopsy showed paranodal axonal swelling, dense masses of neurofilaments and secondary myelin retractions. There was a sudden onset and rapid progression of the disease at a point in time associated with a change in formulation (Table 17) among a population which had been regularly sniffing a thinner for a period of years. The difference in composition was a reduction in the amount of n-hexane and ethyl acetate, an increase in the white spirit fraction and the introducton of MEK. MEK itself is not neurotoxic. The concentrations involved were far in excess of any seen in the workplace. Later, unpublished work suggests no effects at or near the TLV.

TABLE 17

SOLVENT COMPOSITION AND POLYNEUROPATHY

PRODUCT	COMPOSITION, %				
	TOLUENE	n-HEXANE	BENZINE FRACTION	ETHYL ACETATE	MEK
ORIGINAL	30	31	11	28	—
MODIFIED	29	16	26	18	11

FROM ALTENKIRSCH, ET AL. (1977) |n p

Of considerable interest would be the development of data suggesting antagonism by some component of solvent formulations. This might both add to the body of knowledge on mechanism and afford a degree of protection.

Although this paper is a review of recent developments in the toxicity of solvents, some mention of negative studies may also be of value. Studies conducted by the Chemical Industry Institute of Toxicology (USA) on toluene produced no adverse effect (CIIT, 1980 (35)). Rats were exposed for two years to toluene vapor at 300, 100 and 30 ppm on a six hour per day, 5 day per week basis. No exposure-related changes were found using a comprehensive protocol including clinical chemistry and extensive histopathology.

FUTURE RESEARCH NEEDS

Rather than close this review with a summary statement, it might be more valuable to point out areas where future research is needed in order to better define the conditions for the continued safe use of organic solvents. Since so much of what has been done so far and reported in this paper fits the category of "descriptive" toxicology, we are able to report with some conviction that exposure of animals or workers to hydrocarbon, oxygenated and halogenated solvents can result in the development of disorders of the nervous system, liver, kidney and possibly other organs depending on the agent and the dose; that these disorders may be manifest in a number of ways which are non-specific; that the exposure levels required to produce these changes are variable, not well characterized and generally, but not necessarily, in excess of current official limits; and that the structural requirements for most causative agents are unknown. This certainly suggests that the next generation of studies should be characterized as "mechanistic". The need is for understanding what underlies the descriptions. Among such studies one might include those listed in Table 18. In formulating this list, acknowledgement is made of the unpublished report of Gottfries, et al. (1980) (10) and of numerous discussions with colleagues in the U.S. and Europe.

The major dilemma in evaluating the human work is the almost complete absence of information on the actual exposures. It is rare that one can associate a particular clinical outcome with exposure to one or a few defined agents at some defined exposure level. The more troubling the report--either in terms of the potential persuasiveness of the effect or the relatively low level of exposure producing it--the more complex or ill-defined the exposure circumstance. The rich variety of chemical agents falling into the category of "organic solvents" and the unknown and probably unknowable exposure time-concentration profiles are two components of that complexity. Animal studies afford the opportunity to reduce the number of variables but often do not allow the investigation of the many subtle effects (psychomotor performance, psychiatric rating scales) which are of concern for man.

TABLE 18

FUTURE RESEARCH NEEDS—SOLVENTS TOXICOLOGY

GENERAL
- DEVELOPMENT OF WORKER RECORDS SYSTEMS
- DEVELOPMENT OF TEST METHODS, EXPOSURE TRACKING SYSTEMS
- WORKER EDUCATION MODULES

HUMAN
- SCREENING SYSTEMS DEVELOPMENT AND VALIDATION
  — PSYCHOLOGICAL, PSYCHIATRIC, NEUROPHYSIOLOGICAL
  — PHARMACOKINETICS
- INTERACTION STUDIES
  — HORMONES, NEUROTRANSMITTERS, SOLVENTS

ANIMAL
- PHARMACOKINETICS
- BASIC MECHANISM STUDIES
  — CELL DAMAGE, INTERACTIONS
- ADEQUACY OF MODELS

Where to start? It seems that there are several important steps to be taken under the headings of "general", "human studies" and "animal studies". General programs include the support for and encouragement of worker health records systems, including quantitative exposure data and long-term follow-up which will permit the capture of data on actual health experience. Support for methods development in systems management, exposure tracking and recording, test method development (animal and man), and worker education programs is needed. A continuation of the cooperative association of government, academic and industry investigators should maximize the utilization of scarce resources in productive studies rather than adversarial bickering. Although no recommendation is made here for a change in occupational exposure limit to any solvent, the non-specific nature of the effects reported and the absence of exposure data emphasize the need for good surveillance and continual review of the findings.

The human studies needs appear to be in development and validation of test systems (psychological, psychiatric and neurophysiological) to permit screening on a less professional labor intensive basis. It is important to know how much solvent-related disease exists in the workplace and the effectiveness of control measures. Limited, cautious work may be needed in human volunteers exposed to specific chemical entities to validate various organ system functional measurement, especially renal, hepatic and nervous system. Human studies are also needed to complement the relevant animal work on uptake, distribution, storage, biotransformation and excretion of solvents. The important work in the field by Fernandez, et al. (1977) (36) on trichloroethylene, in which mathematical modeling of uptake, excretion and metabolism was checked experimentally, should be more broadly applied. Similar studies have also been conducted by Sato and co-workers (1977) (37), also using trichloroethylene and by Benignus, et al. (1981) (38) on toluene using only rats. Work in animals should build on such metabolic studies with the review of Toftgard and Gustafsson (1981) (39) as a guide. One should also add research on the mechanism of cell damage, interaction between solvents and hormonal and neurotransmitters and interactions between solvents. Finally, some lingering questions on the suitability of animal models for human effects need to be addressed.

In a scientific and medical problem of this magnitude and pervasiveness, one might hope for something more organized than individual investigations or research terms independently laboring away on some aspect of the problem. Some element of cooperative planning, information exchange and shared education could well advance the state of knowledge in this field at a rate which the impatient among us will find more satisfactory.

Thank you for your attention.

REFERENCES

1. Lehmann, K. B. and Flury, F. Toxicology and Hygiene of Industrial Solvents. Williams and Wilkins, Baltimore, 1943. (Chapter 4, p. 34 et seq).
2. Herskowitz, A., Ishii, N. and Schaumburg, H. n-Hexane Neuropathy. A Syndrome Occurring as a Result of Industrial Exposure. New England J. Med. $\underline{285}$: 82-85 (1971).
3. Yamamura, Y. n-Hexane Polyneuropathy. Folia Psychiat. Neurol. Jap. $\underline{23}$: 45-47 (1969).
4. Browning, E. Toxicology and Metabolism of Industrial Solvents. Elsevier, Amsterdam, 1965. (pp. 166 and 204).

5. Kane, L. E., Dombroske, R. and Alarie, Y. Evaluation of Sensory Irritation from Some Common Industrial Solvents. Amer. Ind. Hyg. Assn. J. 41: 451-455 (1980).
6. Hastings, L., Cooper, G. P. and Burg, W. Sensory and Psychomotor Effects of Hydrocarbon Solvents in Humans. Undated draft report to American Petroleum Institute.
7. Lupulescu, A. P. and Birmingham, D. J. Effect of Protective Agent Against Lipid-Solvent-Induced Damages. Arch. Environ. Health 31: 33-36, 1976.
8. Gompertz, D. Solvents - The Relationship Between Biological Monitoring Strategies and Metabolic Handling - A Review. Ann. Occup. Hyg. 23: 405-410 (1980).
9. Brugnone, F., Perbellini, L., Gaffuri, E. and Apostoli, P. Biomonitoring of Industrial Solvent Exposures in Workers Alveolar Air. Int. Arch. Occup. Environ. Health 47: 245-261 (1980).
10. Gottfries, C. G., Knave, B., Sjoberg, L. and Widen, L. Biological Effects of Solvents with Special Reference to the Nervous System. Report from Department of Occupational Health and Karolinska Institute, Stockholm, Sweden, 1980.
11. Cunningham, M. L., Gandolfi, A. J., Brendel, K. and Sipes, I. G. Covalent Binding of Halogenated Volatile Solvents to Subcellular Macromolecules in Hepatocytes. Life Sciences 29: 1207-1212 (1981).
12. Kramer, A., Staulinger, H. and Ullrich, V. Effect of n-Hexane Inhalation on Mono-oxygenase System in Mice Liver Microsomes. Chem.-Brol. Interact. 8: 11-18 (1974).
13. Scala, R. A. Hydrocarbon Neuropathy. Ann. Occup. Hyg. 19: 293-299 (1976).
14. Gustafsson, J. A. Effects of Solvents and Other Industrial Chemicals on Liver Metabolism and Hormone Levels. Report from the Karolinska Institute, Stockholm, Sweden (1980).
15. Schaumburg, H. H. and Spencer, P. S. Toxic Models of Certain Disorders of the Nervous System - A Teaching Monograph. Neurotoxicol. 1: 209-220 (1979).
16. Egan, G. F., Spencer, P., Schaumburg, H., Murray, K. J., Bischoff, M. and Scala, R. A. n-Hexane - "Free" Hexane Mixture Fails to Produce Nervous System Damage. Neurotoxicol. 1: 515-524 (1980).
17. Carpenter, C. P., et al. Petroleum Hydrocarbon Toxicity Studies I - XVII Toxicol. Appl. Pharmacol. 32-41 (1975, 1976, 1977, 1978).

18. Lione, J. G. TSCA Section 8(e) Reports to U.S. EPA. September 21, 1979. (8EHQ-1079-0312) and January 28, 1981.
19. Weaver, N. K. Report of Preliminary Research Findings to U.S. EPA - Unleaded Gasoline Study. November 23, 1981.
20. van der Laan, G. Chronic Glomerulonephritis and Organic Solvents: A Case-Control Study. Int. Arch. Occup. Environ. Health 47: 1-8 (1980).
21. Olsson, H. and Brandt, L. Occupational Exposure to Organic Solvents and Hodgkin's Disease in Men. A Case-Referent Study. Scand. J. Work Environ. Health 6: 302-305, 1980.
22. Juntunen, J., Huphi, V., Hernberg, S. and Luisto, M. Neurological Picture of Organic Solvent Poisoning in Industry; A Retrospective Clinical Study of 37 Patients. Int. Arch. Occup. Environ. Health 46: 219-231 (1980).
23. Seppalainen, A. M. and Lindstrom, K. Solvents and the Nervous System. Lancet II (8251): 864 (October 17, 1981).
24. King, M. D., Day, R. E., Oliver, J. S., Lush, M. and Watson, J. M. Solvent Encephalopathy. Brit. Med. J. 283: 663-665 (September 5, 1981).
25. Altenkirch, H., Mager, J., Stoltenburg, G. and Helmbrecht, J. Toxic Polyneuropathies After Sniffing a Glue Thinner. J. Neurol. 214: 137-152 (1977).
26. Fazio, C. and Ballistini, N. Neuropsychiatric Symptoms in Solvent-Exposed Workers of Shoe Industry. Acta nerv. sup. (Praha) 21: 298-299 (1979).
27. Olson, B. A., Gamberale, F. and Gronqvist, B. Reaction Time Changes Among Steel Workers Exposed to Solvent Vapors. Int. Arch. Occup. Environ. Health 48: 211-218 (1981).
28. Struwe, G., Mindus, P. and Jonsson, B. Psychiatric Ratings in Occupational Health Research: A Study of Mental Symptoms in Lacquerers. Amer. J. Ind. Med. 1: 23-30 (1980).
29. Elofsson, S. A., Gamberale, F., Hindmarsh, T., Inegren, A., Isaksson, A., Johnsson, I., Knave, B., Lydahl, E., Mindus, P., Persson, H. E., Philipon, B., Steby, M., Struwe, G., Soderman, E., Wennberg, A. and Widen, L. Exposure to Organic Solvents: A Cross-Sectional Epidemiologic Investigation on Occupationally Exposed Car and Industrial Spray Painters with Special Reference to the Nervous System. Scand. J. Work Environ. Health 6: 239-273 (1980).

30. Seeber, A. M., Kempe, H. and Schneider, H. Psychological Findings in Solvent-Exposed Workers. Acta nerv. sup. (Praha) 21: 284-285 (1979).
31. Seppalainen, A. M., Lindstrom, K., Martelin, T. Neurophysiological and Psychological Picture of Solvent Poisoning. Amer. J. Ind. Med. 1: 31-42 (1980).
32. Lindstrom, K. Changes in Psychological Performances of Solvent-Poisoned and Solvent-Exposed Workers. Amer. J. Ind. Med. 1: 69-84 (1980).
33. Husman, K. Symptoms of Car Painters with Long-term Exposure to a Mixture of Organic Solvents. Scand. J. Work Environ. Health 6: 19-32 (1980).
34. Husman, K. and Karli, P. Clinical Neurological Findings Among Car Painters Exposed to a Mixture of Organic Solvents. Scand. J. Work Environ. Health 6: 33-39 (1980).
35. Chemical Industry Institute of Toxicology. A Twenty-four Month Inhalation Toxicology Study in Fischer-344 Rats Exposed to Atmospheric Toluene. Executive Summary. Dockett 22000. Oct. 15, 1980 (61 pp.).
36. Fernandez, J. G., Droz, P. O., Humbert, B. E. and Caperos, J. R. Trichloroethylene Exposure. Simulation of Uptake, Excretion and Metabolism Using a Mathematical Model. Brit. J. Ind. Med. 34: 43-55 (1977).
37. Sato, A., Nakajima, T., Fujiwara, Y. and Murayama, N. A Pharmacokinetic Model to Study the Excretion of Trichloroethylene and Its Metabolites After an Inhalation Exposure. Brit. J. Ind. Med. 34: 56-63 (1977).
38. Benignus, V. A., Muller, K. E., Barton, C. N. and Bittikofer, J. A. Toluene Levels in Blood and Brain of Rats During and After Respiratory Exposure. Toxicol. Appl. Pharmacol. 61: 326-334 (1981).
39. Toftgard, R. and Gustafsson, J. A. Biotransformation of Organic Solvents. A Review. Scand. J. Work Environ. Health 6: 1-18 (1980).

MAJOR TOXICOLOGICAL FEATURES OF CHLORINATED SOLVENTS

K.K. Beutel

*Dow Chemical Europe S.A.
Technical Centre
8810 Horgen/Switzerland*

INTRODUCTION

Chlorinated hydrocarbons exist in sizeable numbers, however only a few of them can be used as solvents. Methylene chloride, 1,1,1-trichloroethane, trichloroethylene and perchloroethylene are commonly used as solvents in a variety of industrial applications world wide.

Numerous toxicological studies have been conducted and research is still going on in this field. A complete safety analysis of an operation using chlorinated solvents must include considerations of flammability, boiling point (vapour pressure), vapour density, probability of accidental gross exposure, and of course toxicity.

To show why the above four solvents are popular, chlorinated methanes, ethanes, ethylenes and other chlorinated hydrocarbons are reviewed.

Chlorinated Methanes

Methyl chloride is a flammable gas which is used as a chemical intermediate and as a blowing agent.

Dichloromethane (methylene chloride) is an important solvent used in a variety of applications. The product has a boiling point of 40°C. In most European countries the occupational health standard (TLV) is 100 ppm.

Trichloromethane (chloroform) has a boiling point of 61°C, a TLV of 10 ppm. The product has lost its importance as an industrial solvent, primarily because there are other solvents that are easier to use in a safe fashion. Chloroform has some limited use as a laboratory solvent, however, most is used as a chemical intermediate.

Tetrachloromethane (carbontetrachloride) has a boiling point of 77°C, a TLV of 5 ppm. Carbontetrachloride is no longer used as an industrial solvent. It is primarily used as a chemical intermediate for the production of fluorocarbons.

Chlorinated Ethanes

Some of the chlorinated ethanes are used as chemical intermediates. Important as an intermediate is 1,2-dichloroethane to produce vinyl chloride.

1,1,1-Trichloroethane (methyl chloroform) is the only representative within the group of chlorinated ethanes that has become an important industrial solvent. 1,1,1-Trichloroethane is probably the least toxic of all the chlorinated solvents. The product has a boiling point of 74°C, the TLV was assigned 350 ppm.

1,1,2-Trichloroethane has a boiling point of 114°C, a TLV of 10 ppm, and is considerably more toxic than the 1,1,1-trichloroethane isomer. It has little use as a solvent intermediate.

Tetra- and pentachloroethane have little or no use as a solvent.

Unsaturated Chlorinated Hydrocarbons

Monochloroethylene (vinylchloride) and 1,1-dichloroethylene (vinylidene chloride) are important chemical intermediates as monomers in plastic manufacturing.

Cis- and trans 1,2-dichloroethylene are no longer in use as commercial solvents. Their main disadvantage is their flammability.

Trichloroethylene is still an important commercial solvent. The boiling point is 87°C, the TLV is 100 ppm.

Tetrachloroethylene (perchloroethylene) is also an important commercial solvent, with a boiling point of 121°C, and a TLV of 100 ppm.

Other Chlorinated Hydrocarbons

Within the group of chlorinated propanes, 1,2-dichloropropane (PDC) and 1,2,3-trichloropropane find some use as solvents and as chemical intermediates. Allylchloride and chloroprene are used as chemical intermediates, whereas 1,3-dichloroprene is used as soil fumigant.

Important Industrial Chlorinated Solvents

On the basis of toxicity, flammability and boiling point (vapour pressure), only four chlorinated solvents are important in industrial applications, namely, dichloromethane, 1,1,1-trichloroethane, trichloroethylene and tetrachloroethylene.

Dichloromethane (methylene chloride) is used in aerosol formulations, in paints and paint removers, as an extraction solvent, in adhesives as a urethane foam blowing agent and as a general cleaning solvent.

1,1,1-Trichloroethane (methyl chloroform) is primarily used in metal cleaning. Other applications are in adhesives and coatings, solvent formulations, and miscellaneous uses.

Trichloroethylene is mainly used in metal cleaning, with some limited use in dry cleaning and textile processing.

Tetrachloroethylene (perchloroethylene) is used in drycleaning, textile processing, and metal cleaning. The solvent has some limited use as a chemical intermediate in the production of fluorocarbons.

TOXICITY

All four solvents have similar but varying degrees of effects from oral ingestion, eye, or skin contact. [1]

Each solvent will defat the skin and frequent skin contact may produce dermatitis in certain individuals. Slight irritation will result when the solvent is accidentally splashed into the eye.

Inhalation is by far the most significant route of acute short term exposure in industrial applications.

The first response from acute exposure to excessive amounts of solvent vapour is depression of the central nervous system.
The consequence of exposure to very high vapour concentrations are dizziness and light headedness, followed by loss of consciousness. As concentration and time of exposure increase, death can result if too much vapour is inhaled.

Dichloromethane and 1,1,1-trichloroethane show a lower acute toxicity in comparison to tri- and perchloroethylene.

TABLE I

Acute Toxicity

	LD_{50}*	LC_{50}*	effects CNS humans at 1000 ppm
Methylene Chloride	2g/kg	15000 ppm	20 min
1,1,1-Trichloroethane	10-12g/kg	13500 ppm	60 min
Trichloroethylene	5g/kg	> 3000 ppm	6 min
Perchloroethylene	2.6 g/kg	5000 ppm	2 min

*rats (8 hrs)

Chronic Toxicity

In addition to acute toxicity, it is necessary to fully evaluate repeated and prolonged exposure, mutagenicity, carcinogenicity and teratology. Finally, it is important to understand the metabolism of the solvent.

Mutagenicity and Carcinogenicity

Short-term assays testing the effects of chemicals on bacteria and isolated cells, such as the Ames test and Cell Transformation test, have been used to predict mutagenicity and carcinogenicity. There is considerable debate among knowledgeable scientists about which of these tests have reliable predictive value. While these in vitro tests can be used as indicators for additional work (if testing has not been established on living animals), one should recognize their limitations, namely, the poor qualitative and often variable reproducibility. Certainly there is no way to accurately predict potency based on in vitro tests and thus no way to conduct risk assessments.

Even greater uncertainty is evident if attempts are made to relate the results of these currently available short-term assays to the likelihood that a chemical might cause a mutagenic or carcinogenic response in humans. Clearly when long-term chronic testing has already shown negative results, these take precedence.

Concerning methylene chloride, Ames Tests and Cell Transformation tests have been reported (2, 3) as negative or weakly positive.

Carcinogenic studies have also been completed. (3) Nearly 2000 male and female rats and hamsters were exposed by inhalation to 0, 500, 1500, or 3500 ppm of methylene chloride vapours six hours/day, five days/week for up to two years. The liver was the primary target organ in rats, with slight exposure-related effects in both sexes at 500, 1500, or 3500 ppm. Liver effects included increased hepatocellular vacuolization (consistent with fatty change) in both sexes at 500, 1500, or 3500 ppm, increased multinucleated hepatoxytes in the central lobular region in females exposed to 500, 1500, or 3500 ppm and increased foci and areas of altered hepatocytes in females exposed to 3500 ppm. Liver effects first appeared after 12 months of exposure, progressed slightly from 12 to 18 months, but were unchanged in severity thereafter. There was no increase in hepatocellular carcinomas. A primary target organ was not found in hamsters of either sex. The only effects observed at all levels were primarily the result of decreased amounts of amyloid (a naturally occurring geriatric disease). The number of female rats with a benign tumor did not increase in the exposed versus the control animals, but the total number of benign mammary tumors was increased at the 500, 1500, or 3500 ppm levels. The effect was observed to a lesser extent in male rats, and was only apparent in the 1500 and 3500 ppm groups. This effect was not present in the hamsters of either sex. There was no increase in malignant mammary tumors in any group exposed to methylene chloride. Male rats exposed to 1500 or 3500 ppm appeared to have an increased number of sarcomas (malignant mesenchymal tumors) in the ventral neck region in or around the salivary glands. There were 1, 0, 5, and 11 sarcomas in the 124 male rats per group exposed to 0, 500, 1500, or 3500 ppm, respectively. No increase was observed in these tumors in the female rats or either sex of hamsters. A possible explanation was suggested by the authors: early in their life, these rats suffered from a virus infection in the salivary glands.

The investigators concluded methylene chloride is not likely to be a carcinogen under present or contemplated use conditions.

1,1,1-Trichloroethane has been shown to be negative in most in vitro mutagenic testing. *(4)*

Two lifetime studies have not shown a carcinogenic response: A gavage study in rats was carried out at dose levels of 1500 and 750 mg/kg/day for 2 years *(5)*. More important is an inhalation study on rats exposed to 875 or 1750 ppm of stabilized 1,1,1-Trichloroethane. There was no increase in tumors of any type. *(6)*

Trichloroethylene has been shown to be weakly positive in certain microbial mutagen systems *(7, 8, 9)*, but is negative in higher order studies.

Stabilized trichloroethylene was included in the NCI-Bioassay for carcinogenesis *(10)*. Rats were fed 1000 or 500 mg/kg/day, male mice received 2400 or 1200 mg/kg/day, female mice 1800 or 900 mg/kg/day. Little effect was seen in rats. In the $B_6C_3F_1$ mouse which has a high background incidence of hepatocellular carcinomas, there was an increased incidence in the high dose levels.

Inhalation studies on rats, mice and hamsters at 100 or 500 ppm of pure trichloroethylene did not indicate a carcinogenic potential. *(11)*. It is believed that trichloroethylene does not present a potential problem of carcinogenicity under present or planned use conditions *(1)*.

Perchloroethylene has been shown to be inactive when tested in most in vitro systems. *(12, 13)*

When perchloroethylene was fed by gavage in the NCI bioassay program at 1000 and 500 mg/kg/day to rats and mice, hepatocellular carcinomas were increased in mice but not in rats. *(14)* The high natural occurence of this tumor in the mouse casts severe doubt on the relevance of this result. The carcinogenic effect is claimed to be nongenetic.

An inhalation study done in rats at 600 or 300 ppm of perchloroethylene vapours did not show a carcinogenic response. *(15)*

Teratogenicity and Reproduction

Methylene chloride did not produce a teratogenic effect in rats or mice at vapour concentrations of 1225 ppm *(16)*. Another study on methylene chloride also did not show a teratogenic effect in rats when exposed to 4500 ppm before and during gestation *(17)*.

No adverse effects were noted in a reproduction study when rats were fed 2.8 mg/kg of methylene chloride in the drinking water (125 ppm) for 91 days *(18)*.

Stabilized 1,1,1-trichloroethane was exposed to pregnant female rats and mice at a vapour concentrations of 875 ppm. No effects were observed on the mothers nor on the fetuses *(16)*. Also no effects related to exposure of 2100 ppm were observed in a study involving rats only *(19)*.

1,1,1-Trichloroethane was included in a multigeneration reproduction study. There was no adverse effect in mice. The daily doses in drinking water were 0, 99, 2640 and 8520 mg/kg *(20)*.

A trichloroethylene study was carried out at 300 ppm in rats and mice with no indication of a teratogenic effect *(16)*. Another study, on rats only, confirmed these negative results at 1800 ppm *(21)*.

A perchloroethylene study was also carried out at 300 ppm in rats and mice. There was no teratogenic effect observed *(16)*. Two additional studies, one at 500 ppm and the other at 1000 ppm showed no teratogenic effects in rats *(22, 23)*.

Metabolism / Pharmacokinetics

Absorption and excretion of methylene chloride has been studied extensively over the last years *(24)*. Special emphasis was placed on the metabolism to form carboxyhemoglobin.

Humans that were exposed to 100 ppm methylene chloride for 8 hrs had a CoHb value of 3.2 %, whereas an 8 hrs exposure at 150 ppm produced 5.4 % and an 8-hour exposure at 200 ppm resulted in CoHb level of 6.8 % *(25)*.

Five % CoHb corresponds to a 35 ppm exposure for carbon monoxide.

The low toxicity of 1,1,1-trichloroethane may be related to the small amount of metabolism which occurs in animals and man. Only small quantities of trichloroacetic acid and trichloroethanol are found in the urine, most of the solvent vapour (97.6 %) is exhaled unchanged *(26)*.

Trichloroethylene on the other hand metabolizes easily; trichloroacetic acid and trichloroethanol can be found in the urine. The metabolism may take place through an epoxy intermediate and chloral hydrate to trichloroacetic acid or trichloroethanel. High blood ethanol concentrations interfere with the metabolism *(27)*.

Perchloroethylene appears to be less metabolized than trichloroethylene, yet its hepatotoxicity seems to be related to metabolism. Only limited metabolism takes place in humans. At 50 ppm only 20 % of the metabolism occurs compared to trichloroethylene *(28)*.

Health and Epidemiological Studies

A retrospective mortality study of a methylene chloride exposed population showed no solvent related increase in death due to any specific cause. Over 30 % of the exposed population had worked a minimum of 20 years by 1964. In addition to close health follow-up of all methylene chloride exposed persons from that time on, the long-term segment of the population represented occupational exposure of over 30 years ranging from 30 to 125 ppm TWA. There was no indication of solvent-related carcinogenicity, or cardiovascular effects *(29)*.

A rather extensive health study of 151 men and women has been reported. The workers had been exposed to 1,1,1-trichloroethane for several months up to six years. During the study period exposures for some workers exceeded 200 ppm. Based on subjective responses and some previous monitoring data, exposure concentrations had been higher prior to the study period. When compared to 151 matched pair control subjects by numerous medical and physiological parameters, there were no adverse effects related to exposure *(30)*.

For trichloroethylene and perchloroethylene epidemiologic studies have been of limited size and value.

In one study the cancer mortality of trichloroethylene workers was investigated, involving 518 men with rather low levels of exposure. No excess cancer mortality was reported *(31)*.

A preliminary analysis of death certificates of 330 former dry-cleaning workers was reported. Neither cancer cases nor noncancer cases were associated with years of working with perchloroethylene *(32)*.

INDUSTRIAL HYGIENE STANDARDS

The Threshhold Limit Values (TLV's) of the American Conference of Governmental Industrial Hygienists (ACGIH) refer to airborne concentrations under which it is believed that nearly all workers may be repeatedly exposed day after day without adverse effect *(33)*. The TLVs are also adopted by a number of European countries. The German MAK-values, set by the Commission for Investigation of Chemical Compounds in the Work Area have a similar definition as the TLVs.

In general the occupational hygienic standards for these solvents are lower in Eastern Europe and in Scandinavia.

For methylene chloride the TLV time-weighted average

concentration as well as the MAK-value were lowered to 100 ppm on the basis of increased carboxyhemoglobin concentration. The 15 minute-Short Term Exposure Limit (TLV-STEL) for methylene chloride was recommended to be 500 ppm since neither undesirable CNS responses nor excessive carboxyhemoglobin values are expected with such short exposure. *(34)*

On the basis of its low toxicity 1,1,1-<u>trichloroethane</u> has a TLV of 350 ppm to prevent beginning anesthetic effects and objections to odour. The 15 minute Short Term Exposure Limit of 450 ppm is recommended for protection against anesthesia. To increase the safety margin, the German MAK-value was set at 200 ppm.

TABLE II

Industrial Hygiene Standards

	TLV/ACGIH (1981) TWA	STEL (15 min)	MAK (1981) W.Germany	BASIS
Methylene chloride	100 ppm	500 ppm	100 ppm	carboxy-hemoglobin
1,1,1-Trichloro-ethane	350 ppm	450 ppm	200 ppm	odour
Trichloroethylene	100 ppm (50 ppm)*	150 ppm	50 ppm**	CNS
Perchloroethylene	100 ppm (50 ppm)*	150 ppm	100 ppm	CNS

* list of intended changes   ** carcinogen class III B

At present the TLV for <u>trichloroethylene</u> is 100 ppm, however it is listed in the notice of intended changes for 1981 with a recommendation of 50 ppm. The MAK-value was set at 50 ppm. Trichloroethylene is also listed under "Carcinogenic working materials" section B. The paragraph lists those chemicals for which a considerable carcinogenic potential can be suspected and which urgently need further clarification *(35)*.

Trichloroethylene is not listed as a suspect carcinogen in the ACGIH-TLV book.

The TLV for perchloroethylene is presently 100 ppm, with an intended change of 50 ppm. In the 1981-list the MAK-value was set at 100 ppm also with the proposal to reduce the standard to 50 ppm in 1982.

CONCLUSION

The toxicity of the four major chlorinated solvents has been thoroughly studied.

The first response from acute exposure to excessive amounts of solvent vapours is depression of the central nervous system.

In general, it can be stated that, within prescribed handling and use procedures, these four chlorinated solvents can be used without harm to people occupationally exposed to their vapours.

Extensive testing and human experience indicate that methylene chloride and 1,1,1-trichloroethane have a favoured position; 1,1,1-trichloroethane appears to be the least toxic of all the chlorinated solvents.

REFERENCES

1. T.R. Torkelson and V.K. Rowe, *Industrial Hygiene and Toxicology*, 3rd ed., Vol 2B, ed. George D & Florence Clayton, 1981.
2. W.M.F. Jongen, G.M. Alink, and J.M. Koemann, *Mutat. Res.*, 56, 245 (1978).
3. Dow Chemical U.S.A., *Methylene Chloride: A Two-Year Inhalation Toxicity and Oncogenicity Study in Rats and Hamsters*. Unpublished report of work done for Diamond Shamrock Corporation, Dow Chemical U.S.A., Imperial Chemical Industries, Ltd. (UK), Stauffer Chemical Company, and Vulcan Materials Company by the Toxicology Research Laboratory, Dow Chemical U.S.A., Midland, MI 48640, December 31, 1980.
4. The Dow Chemical Company, Midland, Mich., unpublished data.

5. National Cancer Institute, *Bioassay of 1,1,1-Trichloroethane for Possible Carcinogenicity*, NCI-CG-TR-3 (1977).
6. J.F. Quast, L.W. Rampy, M.F. Balmer, B.K.J. Leong, and P.J. Gehring, *Toxicologic and carcinogenic evaluation of a 1,1,1-trichloroethane formulation by chronic inhalation in rats*. Toxicology Research Laboratory, Dow Chemical U.S.A., Midland, MI 48640, October 6, 1978.
7. L. Fishbein, *Mutat. Res.*, 32, 267 (1976).
8. H. Greim, G. Bonze, Z. Radwan, D. Reichert, and D. Henschler, *Biochem. Pharmacol.*, 24, 2013 (1975).
9. G. Bronzetti, E. Zeiger, and D. Frezza, *J. Environ. Phatol. Toxicol.*, 1, 411 (1978).
10. National Cancer Institute, *Bioassay of Trichloroethylene for Possible Carcinogenicity*, NCI-CG-TR-2, 1976.
11. D.W. Henschler, W. Roman, H.M. Elsässer, D. Reichert, E. Eder, and Z. Radwan, *Arch. Toxicol.*, 43, 237 (1980).
12. H. Greim, D. Bimboes, G. Egert, W. Goeggelmann, and M. Kraemer, *Arch. Toxicol.*, 39, 159 (1977).
13. H. Bartsch, C. Malaville, A. Barbin, and G. Planche, *Arch. Toxicol.*, 41, 249 (1979).
14. National Cancer Institute, *Bioassay of Tetrachloroethylene for Possible Carcinogenicity*, DHEW Publication No. (NIH)77-813, 1977.
15. L.W. Rampy, J.F. Quast, B.K.J. Leong, and P.J. Gehring, *Proceedings of the First International Congress on Toxicology*, G.L. Plaa and W.A.M. Duncan, Eds., Academic Press, New York, 1978, p. 562.
16. B.A. Schwetz, B.K.J. Leong, and P.J. Gehring, *Toxicol. Appl. Pharmacol.*, 32, 84(1975).
17. R.D. Hardin and J.M. Manson, *Toxicol. Appl. Pharmacol.*, 52, 22 (1980).
18. G. Bornmann, and A. Loesser, *Z. Lebsnmittel-Untersuch Forsch.* 136, 14(1967), through FAO Nutrition Meetings Report Series No. 48A WHO/FAO Food Add/70.39 (1970).
19. I.J.G. Climie, D.H. Hutson, B.J. Morrison, and G. Stoydin, *Xenabiotica* 3, 149 (1981).
20. B.L. Riddle, R.A. Carchman, and J.F. Borzelleca, *The Toxicologist* 1, 26 (1981).
21. Evaluation of teratogenicity and behavioral toxicity with inhalation exposure of maternal rats to trichloroethylene. MA. Dorfmueller, SP. Henne, RG. York, RL. Bornschein, JM. Manson. *Toxicology*, Vol. 14, (1979).
22. Behavioral Teratology of perchloroethylene in rats. BK. Nelson, B.J. Taylor, JV. Setzer, R.W. Hornung. Journal of Environ. *Pathol. and Toxicol.* Vol. 3, (1980).

23. Teratogenic-mutagenic risk of workplace contaminants. Trichloroethylene, perchloroethylene and carbon disulfide. RP, Beliles, DJ. Brusick and FJ. Mecler. *Litton Bionetics Report* NIOSH Contract No. 210-77-0047, May (1980).
24. M.J. McKenna, J.H. Saunders, W.H. Boeckler, R.J. Karbowski, K.D. Nitschke, and M.B. Chenoweth, Abst. 176, *Soc. Toxicol. 19th Ann. Meet. March 9-13, 1980.*
25. DiVincenzo, G.F. Yanna, F.J. Astill, B.D., *Am. Ind. Hyg. Assoc. J.* 33:125 (1972).
26. C.L. Hake, T.B. Waggoner, D.N. Robertson, and V.K. Rowe, *Arch. Environ. Health*, 1, 101 (1960).
27. L.L. Miller, *Occup. Med.*, 5, 194 (1948).
28. M. Ikeda, *Environ. Health Perspect.*, 21, 239 (1977).
29. B.R. Friedlander, T. Hearne, and S. Hall, *J. Occup. Med.*, 20, 657 (1978).
30. C.G. Kramer, M.G. Ott, J.E. Fulkerson, N. Hicks, and H.R. Imbus, *Arch. Environ. Health*, 33- 331 (1978).
31. O. Alexson, K. Andersson, C. Hogstedt, B. Holmberg, G. Molinz, and A. De Verdier, *J. Occup. Med.*, 20, 194 (1978).
32. A. Blair, P. Decoufle, and D. Grauman, *Am. J. Publ. Health*, 69, 508 (1979).
33. Threshold Limit Values for Chemical Substances in Workroom Air. *Adopted by ACGIH* for 1981.
34. Documentation of the Treshold Limit Values. *American Conference of Governmental Industrial Hygienists*, Cincinnati (1981).
35. Maximale Arbeitsplatzkonzentrationen 1981, Krebserzeugende Arbeitsstoffe III B. H. Bolt Verlag (1981).

SOME ASPECTS OF THE COMPARATIVE TOXICOLOGY OF
CHLORINATED SOLVENTS

A M Moses

Division Medical Department
I.C.I. P.L.C., (Mond Division)
P.O. Box 13, The Heath,
Runcorn, Cheshire WA7 4QF, UK

1. INTRODUCTION

The chlorinated solvents considered in this paper are: trichloroethylene, perchloroethylene, 1,1,1-trichloroethane and methylene chloride.

These four solvents have been the subjects of a very considerable amount of investigatory work, both human and animal, and in consequence the data base available is probably as extensive as that for any other group of solvents.

Within this mass of data, which I cannot attempt to review in detail in the time available, it is inevitable that some aspects and some solvents have been more thoroughly investigated than others. What I would like to do is to refer in broad terms to some of the areas of study which I believe to be most relevant to an overall assessment of the toxicology (and in particular to the establishment of hygiene standards), indicating where the weight of investigation has been and giving some views on the scientific difficulties facing both the authors and the interpreters of some of these studies.

2. PRESENTATION

The areas I would like to review are the following:

(a) Animal Toxicity.

Specifically repeated administration by the inhalation route.

(b) Animal Carcinogenicity.

(c) Metabolism.

(d) Human Studies, subdivided into:

   i. Case reports of poisoning (acute and chronic).
   ii. Volunteer studies.
   iii. Field studies and epidemiology.

(a) Animal Toxicity

Several classical subacute experiments by the inhalation route appear in the literature. For trichloroethylene (1), the picture is one of slight increase in liver weight with minimum histopathological change, along with slight depression of growth. No effect levels are 400 ppm for the monkey, 200 ppm for rats and rabbits and 100 ppm for the guinea pig. Perchloroethylene presents a similar picture (2), but with rather more marked histopathological changes in the liver, although no effect levels are 400 ppm for rat, rabbit and monkey, with the guinea pig showing a minimal effect at 100 ppm. With 1,1,1-trichloroethane the position is rather different (3), with virtually no effect on the liver; growth rate depression is seen at levels above 3,000 ppm in rats, rabbits and monkeys, with a minimal effect on guinea pigs at 650 ppm. An old study on methylene chloride (4) indicates 500 ppm to be without effect after six months exposure in dogs, rabbits and rats, although guinea pigs showed adverse liver effects. The apparent sensitivity of the guinea pig to these solvents may be related to a different balance between activating and detoxifying systems.

All these studies involve an exposure regime broadly corresponding to occupational exposure (ie 6 or 7 hr/day, 5 days/week). There are in addition a number of published studies under continuous exposure regimes; those studies, while relevant to specific situations, are much more difficult to interpret in terms of occupational exposure.

(b)   Animal Carcinogenicity

Trichloroethylene has been found to be positive (ie to produce liver tumours) following high level oral dosage in the NCI mouse, but not in the NCI rat (5). It has also been found negative by Maltoni (6) (using ingestion in the rat), Henschler (7) (using inhalation in the rat, mouse and hamster) and van Duuren (8) (using skin routes in the mouse). Perchloroethylene, while again producing liver tumours following high oral doses in the NCI mouse, was negative in the NCI rat (9), and also negative in an inhalation study (10) in the rat. 1,1,1-trichloroethane did not produce any increased tumour incidence in either the NCI mouse or the NCI rat (11), and was also negative in two inhalation studies (13,14) although all three studies have been criticised. Methylene chloride has been shown, at high doses by the inhalation route, to be negative in the hamster, while producing an increased incidence of malignant salivary gland tumours at the top dose in the male rats only (15).

Our view is that, in the light of all the available data (including epidemiology) none of these solvents is likely to present a carcinogenic risk to humans when handled in accordance with good occupational practice.

(c)   Metabolism

Trichloroethylene is believed to pass through an epoxy intermediate and chloral hydrate to trichloroacetic acid and trichloroethanol. The actual percentage metabolised depends on dose and route of absorption, but there are reports indicating that more than 50% of a low inhalation dose may be metabolised. Perchloroethylene also appears to proceed via an epoxide to trichloroacetic acid, with trichloroethanol and a number of other species being mentioned. The extent of metabolism is rather less than for trichloroethylene, although it is difficult to be quantitative and there are species differences. 1,1,1-trichloroethane is metabolised by animals and man to a very small extent, and only small amounts of trichloroacetic acid and trichloroethanol are found in the urine.

The low systemic toxicity of 1,1,1-trichloroethane may be related to this small extent of metabolism. The metabolism of methylene chloride to carbon monoxide and thence carboxyhaemoglobin has been thoroughly investigated and needs no further comment here.

(d) Human Studies

(i) Case Reports. There have been a number of cases of trichloroethylene poisoning reported, including some fatalities (16,17,18). The predominant effect is that on the central nervous system, with several fatalities being ascribed to ventricular fibrillation. There have also been a number of reports relating liver damage to trichloroethylene over-exposure (18). In the majority of these a clear causative link is not established and they must be viewed against the background that, in the past, trichloroethylene undoubtedly contained significant quantities of hepato-toxic impurities.

There are several case reports of deaths from anaesthetic effects following over-exposure to perchloroethylene (19,20); cardiosensitisation does not appear to have been a problem. Liver injury following repeated excessive exposure has been reported in some subjects and we believe this to be rather better substantiated than in the case of trichloroethylene.

Several cases of over-exposure to 1,1,1-trichloroethane have been reported, some with fatal outcome (21,22,23). There are also reports of sudden deaths following 'sniffing' incidents, where cardiosensitisation has been implicated (22). There is little suggestion of any liver effect and uneventful recovery of non-fatal cases is a feature of these reports.

In the case of methylene chloride, there are a few reports of fatal accidents, each the result of gross over-exposure (24). Recovery appears to be uneventful from non-fatal cases.

These reports are of limited direct value in the establishment of a hygiene standard, but they do provide useful background to an overall assessment of the toxicology.

(ii) Volunteer Studies. There are a considerable number of reports of studies involving human volunteers. For trichloroethylene, eight such studies (25-32) are referred to. The parameters investigated generally centre on subjective CNS responses, along with psychophysiological tests. Exposure levels vary from 30 to 1,000 ppm. Such studies inevitably concern themselves only with immediate responses, but as such they are clearly relevant in the context of hygiene standards. A fundamental difficulty with such work is that psychophysiological investigations should be carried out to a double-blind protocol, and the smell threshold for trichloroethylene is low enough to make this very difficult (in only one of these studies is this point recognised in the experimental design, although several authors recognise the complicating influence of a knowledge of exposure). Overall, these studies do not indicate an adverse effect at or below 100 ppm, although there are some conflicting reports.

For perchloroethylene, Hake and Stewart (33) have reviewed their detailed studies over the years, including assessment of neurological, physiological, behavioural and subjective responses and conclude that 100 ppm is probably without effect. Again, the difficulties attending a double-blind protocol will have been a complicating factor.

There are several reports of investigations of psychophysiological and subjective responses following exposure to 1,1,1-trichloroethane (34-37). Once more, the absence of a double blind protocol makes interpretation difficult, but the general picture is of the absence of adverse effects at or below 350 ppm.

With methylene chloride, Stewart (40) reported no untoward subjective or objective health response following repeated exposure to 250 ppm. With this solvent, attention has focussed largely on the formation of carboxyhaemoglobin and two recent studies in this area (38,39) have determined carboxyhaemoglobin levels in the blood and derived pharmacokinetic models.

(iii) Field Studies and Epidemiology. Human field studies I define as being studies of the health status of groups of people exposed because of the nature of their occupation to the solvent in question. I have included epidemiological studies under this heading.

These field studies generally include clinical observations, liver function tests, an assessment of subjective responses (usually relating to CNS and gastric symptoms) and in some cases psychophysiological tests. Because of their very nature (ie observations made in an actual occupational situation), these studies should clearly be very relevant to the establishment of hygiene standards; inherent in their nature also are considerable difficulties facing both the investigator and the interpreter.

Because of their relevance, these studies merit close critical examination in this light. Among the problems inherent in these studies are:

the difficulty of defining adequate control populations,

the need to determine exposure levels by an adequate method (eg personal monitor),

difficulties in estimating historical (as opposed to contemporary) exposure levels,

the subjective nature of many of the symptoms reported.

Seven field studies are listed for trichloroethylene (26,28,41-45). The authors of these studies acknowledge to varying degrees the inherent difficulties, and it is not surprising that there are several inconsistencies between studies. Subjective prenarcotic symptoms predominate in the reports, and it is extremely difficult to relate the incidence of these symptoms in a meaningful way to actual exposure levels. Although several of the studies report atmosphere levels below 100 ppm, the inherent difficulties attending these investigations make it difficult to sustain the view that an adverse effect on health at levels below 100 ppm has been demonstrated. In my view this might well prove to be a fruitful area for future investigations.

Three similar studies are listed for perchloroethylene. (46-48). The studies reported neurological, behavioural and subjective symptom assessments, as well as liver function tests. Again, difficulties with controls and atmosphere monitoring make interpretation difficult, but these studies provide little evidence of an adverse effect at or below 100 ppm.

Five reports of field studies on 1,1,1-trichloroethane workers are listed (49-53). The reports cover neurophysiological, behavioural, clinical and general health surveys. No adverse effect has been established at levels below 500 ppm.

There are very few field studies as such on methylene chloride; one such (54) investigates carboxyhaemoglobin levels at atmosphere levels in excess of 100 ppm.

Good epidemiological studies are few and far between. On trichloroethylene, the study by Axelson (55) on 518 Swedish workers shows no increased cancer mortality, although covering a fairly short elapsed time. The studies by Malek et al. (56,57) provide limited evidence that occupational exposure to trichloroethylene is not associated with an increased incidence of liver cancer.

No meaningful studies have been published on perchloroethylene.

On 1,1,1-trichloroethane, Kramer (58) reported a matched pair study on two adjacent textile plants. Numerous physiological parameters were measured and the authors concluded that no health impairment was suffered by the workers at the concentrations measured (a mean of 115 ppm). The limited duration of exposure (up to 6 yrs) to levels which are low relative to the TLV limit the value of this study.

In the case of methylene chloride a mortality analysis of 750 workers exposed to the substance for up to 30 years has been reported (59). Proportionate mortality analysis revealed no significant excess cause of death and life expectancy was equal to or better than the 3 control populations. Exposure levels quoted range around 100 ppm. In our view this is a well conducted study, providing considerable reassurance.

3. CONCLUSION

In conclusion, I would like to make a plea that hygiene standards for these and other substances be based in a rational way on an indepth, critical review of the various studies available. Such a review would include an analysis, not just of the relevance of the study, but of the scientific difficulties inherent in the work and attending interpretation. While the process of standard setting in this and several other countries follows these principles, I feel that there are standard setting bodies around the world where a more critical approach might be adopted. Animal studies, while entailing the overriding problem of extrapolation to man, should be objective and control of variables, while not easy, may be less intractable than in human studies. Human studies present very considerable scientific problems and, while they should clearly be more relevant, these scientific difficulties may limit their value.

For these four solvents, I would like to summarise the main features of the toxicology as follows. For all four solvents, the most serious aspect is likely to be acute over-exposure. This will lead to CNS effects which can prove fatal in extreme cases. In addition several of them have been associated with sudden fatalities at high exposure levels following cardiosensitisation. Repeated over-exposure to perchloroethylene may give rise to adverse liver effects, but this is regarded as unlikely with the others.

With all four, we do not believe that any adverse effects on health are likely to ensue provided they are handled in accordance with the manufacturers instructions which will include observation of the currently prevailing hygiene standards.

REFERENCES

1. Adams, E. M. et al. (1951). Arch. Ind. Hyg. Occup. Med. 4, 469-481.

2. Rowe, V. K. et al. (1952). Arch. Ind. Hyg. Occup. Med. 5, 566-579.

3. Adams, E. M. et al. (1950), Arch. Ind. Hyg. Occup. Med. 1, 225-236.

4. Heppel, L. A. et al. (1944). Ind. Hyg. Toxicol. 26, 8-16.

5. National Cancer Institute. (1976). Carcinogen Tech. Rep. Ser. No. 2. NCI-CG-TR-2, DHEW (NIH) 76-802.

6. Maltoni, C. (1977). Gli Ospedali della Vita, 4, 108-110.

7. Henschler, D. (1980). Arch. Toxicol. 43, 237-248.

8. van Duuren, B. L. (1979). J. Nat. Cancer Inst. 63, 1433-1439.

9. National Cancer Institute. (1977). "Bioassay of tetrachloroethylene for possible carcinogenicity." DHEW Publication No. NIH 77-813.

10. Rampy, L.W. (1978). Unpublished study by Toxicol. Research Lab., Dow Chemical Company, Midland, Michigan. Cited in 12.

11. National Cancer Institute. (1977). "Bioassay of 1,1,1-trichloroethane for possible carcinogenicity." NCI-CG-TR-3.

12. Environmental Protection Agency. (1979). "An Assessment of the Need for Limitations on Trichloroethylene, Methyl Chloroform and Perchloroethylene." Final Report No. EPA-560/11-79-009.

13. Quast, J. F. (1978). Report HET K-1716-56. Dow Chemical Company. Cited in 12.

14. Bell, Z. G. (1978). PPG Industries Inc. Letter to S C Mazaleski, EPA. Cited in 12.

15. Dow Chemical. (1980). "Methylene chloride: A two year inhalation toxicity and oncogenicity study in rats and hamsters". FY1-OTS-0281-0097. Cited in "Assessment of Testing Needs: Dichloromethane". EPA 560/2-81-003, May 1981.

16. Stuber, K. (1932). Arch. Gewerbepathol. Gewerbehyg. 2, 398-456.

17. Kleinfeld, M. (1954). Arch. Ind. Hyg. Occup. Med. 10, 134-141.

18. NIOSH. (1973). "Criteria for a Recommended Standard - Occupational Exposure to Trichloroethylene." US DHEW, HSM 73-11025.

19. von Oettingen, W. F. (1955). In "The Halogenated Hydrocarbons, their Toxicity and Potential Dangers." US Public Health Serv. Publ. No. 414.

20. NIOSH. (1976). "Criteria for a Recommended Standard - Occupational Exposure to Tetrachloroethylene." US DHEW Publ. No. (NIOSH) 76-185.

21. Stewart, R. D. and Andrews, J. T. (1966). J. Amer. Med. Assoc. 195, 904-906.

22. Bass, M. (1970). J. Amer. Med. Assoc. 212, 2075-2079.

23. NIOSH. (1976). "Criteria for a Recommended Standard - Occupational Exposure to 1,1,1-trichloroethane". US DHEW Publ. No. (NIOSH) 76-184.

24. Torkelson, T. R. and Rowe, V. K. (1981). In Patty's "Industrial Hygiene and Toxicology." Vol. 2B, (Eds G. D. Clayton and F. E. Clayton). p.3458. Wiley-Interscience, New York.

25. Ertle, T. et al. (1972). Arch. Toxicol. 29, 171.

26. Konietzko, H. (1979). Fortschr. Med. 97, 671-674.

27. Nakaaki, K. et al. (1973). Rodo Kagaku. 49, 499 (Chem. Abstr. 80, 56210, (1974)).

28. Nomiyama, K. and Nomiyama, H. (1977). Int. Arch. Occup. Environ. Health. 39, 237-248.

29. Salvini, M. et al. (1971). Br. J. Industr. Med. 28, 293-295.

30. Stewart, R. D. et al. (1970). Arch. Environ. Health, 20, 64-71.

31. Stopps, G. J. and McLaughlin, M. (1967). Am. Ind. Hyg. Assoc. J. 28, 43-50.

32. Vernon, R. J. and Ferguson, R. K. (1969). Arch. Env. Health. 20, 462.

33. Hake, C. L. and Stewart, R. D. (1977). Environ. Health Perspect. 21, 231.

34. Gamberale, F. et al. (1972). Arbete. Och. Halsa. 1, 29-50.

35. Salvini, M. et al. (1971). Br. J. Indust. Med. 28, 286-292.

36. Stewart, R. D. et al. (1961). Am. Ind. Hyg. Assoc. J. 22, 252-262.

37. Torkelson, T. R. et al. (1958). Am. Ind. Hyg. Assoc. J. 19, 353-362.

38. Di Vincenzo, G. D. and Kaplan, C. J. (1981). Tox. Appl. Pharmac. 59, 130-140.

39. McKenna, M. J. et al. (1980). Abstr. 176, Soc. Toxicol. 19th Ann. Meet. March 9-13, 1980.

40. Stewart, R. D. et al. (1974). "Methylene Chloride, Development of a Biological Standard." Medical. Coll. of Wisconsin, Milwaukee. NIOSH-MCOW-ENVM-MC-74-9.

41. Ahlmark, A. and Forssman, S. (1951). Arch. Ind. Hyg. Occup. Med. 3, 386.

42. Ahlmark, A. and Friberg, L. (1955). Nordisk Hyg. Tidskrift. 36, 165.

43. Bardodej, Z. and Vyskocil, J. (1956). Arch. Ind. Health. 13, 581.

44. Grandjean, E. et al. (1955). Br. J. Industr. Med. 12, 131-142.

45. Lilis, R. et al. (1969). Med. d. Lavoro. 60, 595.

46. Franke, W. and Eggeling, F. (1969). Med. Welt. 9, 453-460.

47. Munzer, M. and Heder, K. (1972). Zentralbl. Arbeits. Med. 22, 133-138.

48. Tuttle, T. C. et al. (1976). "Behavioural and Neurological Evaluation of Workers exposed to Perchloroethylene". Cited in 20.

49. Binaschi, S. et al. (1969). Boll. Soc. Ital. Biol. Sper. 45, 94-96.

50. Hatfield, T. R. and Maykoski, R.T. (1970). Arch. Environ. Health. 20, 279-281.

51. McConnell, J. C. and Schiff, H. I. (1978). Science. 199, 174.

52. Stewart, R. D. (1963). J. Occup. Med. 5, 259-262.

53. Stewart, R. D. (1968). Ann. Occup. Hyg. 11, 71-79.

54 Ratney, R. S. et al. (1974). Arch. Environ. Health 28, 223-226.

55. Axelson, O. et al. (1978). J. Occup. Med. 20(3), 194-196.

56. Novotna, E. et al. (1979). Prak. Lek. 31(4), 121-123.

57. Malek, B. et al. (1979). Prak. Lek. 31(4), 124-126.

58. Kramer, C. G. et al. (1978). Arch. Environ. Health. 33, 331.

59. Friedlander, B. R. et al. (1978). J. Occup. Med. 20, 657.

CARCINOGENICITY OF SOLVENTS

P. Grasso

Group Occupational Health Centre, BP Research Centre, Chertsey Road, Sunbury-on-Thames, Middlesex TW16 7LN, UK

INTRODUCTION

Comparatively few solvents have been tested adequately for carcinogenicity in animals. Some have given no indication of carcinogenic potential (Table 1) but others have given rise to tumours in different organs of experimental animals (Table 2).

TABLE 1

Solvents tested for carcinogenicity with negative results

Solvent	Route of administration	Species	Ref.
Acetone	Skin (topical)	Mouse	12
"	Vaginal (topical)	"	13
Cyclohexane	Skin	"	14
1,1-Dichloroethane	Oral	Rat	15
Ethylene glycol	Oral	"	16
Propylene glycol	Oral	"	16
"	Subcutaneous	"	17
"	Oral	Mouse	18
"	Skin	"	19
Diethylene glycol monoethylether	Oral	Rat	16
Ethanol	Oral	"	20
Styrene	Oral	Mouse & rat	21
Toluene	Skin (topical)	Mouse	22
Trichloroethylene	Oral	Mouse & rat	23

TABLE 2

Solvents tested for carcinogenicity with positive results

Solvent	Route of administration	Species	Site of carcinogenic action	Ref.
Carbon tetrachloride	Oral, inhalation & subcutaneous	Mouse, rat & hamster	Liver	24
Chloroform	Oral	Mouse & rat	"	24
1,4-Dioxane	"	Rat	"	23
Hexachloroethane	"	Mouse	"	25
1,1,2-Trichlorethane	"	"	"	25
1,1,2,2-Tetrachloroethane	"	"	"	25
Tetrachloroethylene	"	"	"	25
1,2-Dichloroethane	"	Rat	Forestomach & mammary glands	26
"	"	Mouse	Mammary glands	26
Formaldehyde	Inhalation	Rat	Upper respiratory tract	27
"	Subcutaneous	"	Subcutaneous tissue	10
Epichlorhydrin	Inhalation	"	Upper respiratory tract	28
Diethylene glycol	Oral	"	Bladder	7

The conclusion arrived at by the original authors in the majority of instances where the response was a positive one was that the compounds possessed carcinogenic activity. Although this conclusion is valid in terms of animal experimentation one cannot deduce that all of these compounds present a carcinogenic risk to man.

In general, the validity of animal studies in assessing the carcinogenic potential of a chemical to man rests on the demonstration that out of 18 chemicals and processes known to be carcinogenic to man, all but two (benzene and arsenic) have been shown to be carcinogenic in animals (1). Many scientists in this field assume that any chemical which produces tumours in animals, irrespective of the dose level or the route of administration used, is also likely to present a carcinogenic hazard to man. While one accepts the general validity of this assumption, one cannot ignore the evidence that certain types of tumours in animals are less meaningful in this respect than others.

From a review of the results of carcinogenicity tests on solvents, mainly in mice and rats (Table 2), it can be seen that the tumours produced are most commonly found in the liver, upper respiratory tract, mammary glands, bladder and forestomach. There are grounds for suspecting that these tumours arise either because of the genetic predisposition of the animal, inappropriate route of administration of the chemical or some similar reason which leads one to infer that such tumours have little validity on which to base a judgement regarding the carcinogenic hazard to man.

The basis for making this statement with respect to each of the tumours mentioned is now discussed.

Upper respiratory tract tumours

The upper respiratory tract of the rat, like that of most mammals, is covered by a specialized type of epithelium consisting of columnar cells, some of which secrete mucus and others possess cilia whose main function is to keep the mucus constantly on the move. This epithelium is easily damaged by irritants. If the exposure to the irritant is of short duration then the damage is easily repaired and the epithelium is restored to normal, but if the exposure is prolonged, then the damage is repaired by the development of a different type of epithelium at the site of injury. The new epithelium resembles that of normal skin and is called squamous epithelium. The change over from one type of epithelium to another is termed metaplasia and is usually regarded as indicative of serious injury to the tissues. There is some evidence that an injury of this type renders the epithelium susceptible to the development of cancer (2). This evidence came to light recently in an attempt to evaluate the carcinogenicity of formaldehyde vapour (2). Several carcinomas were produced by exposing rats for several months to this vapour at a concentration which produced squamous cell metaplasia. At concentrations which did not do so no tumours were produced. In contrast, when alkylating agents such as dimethyl sulphate are administered by inhalation, they produce not only squamous cell carcinomas but also other types of tumours which originate from the cells normally present in the nasal cavity (3). Hence the induction of squamous cell carcinoma by the solvents listed in Table 2 does not constitute valid evidence to suggest that human beings are likely to develop cancer from exposure to these materials.

Mammary gland tumours

These tumours occur spontaneously in both rats and mice but there is a marked difference in the susceptibility of different strains to the development of these tumours, especially in the rat (4). Certain strains, particularly the Sprague-Dawley, develop a very high natural incidence; other strains do not. The reason for this high natural incidence is not at all clear but some genetic or hormonal factor is suspected.

In the mouse, most strains have a high natural incidence. There is a considerable amount of evidence which shows that this high incidence is due to a combination of three factors: the presence of a virus, a hormonal factor and a genetic factor (5).

Many chemical agents cause an increased incidence of mammary tumours in both rats and mice (5) but the tendency among some regulatory authorities is to discount an increase in the incidence of such tumours in animals as evidence of carcinogenicity in man (6).

Bladder tumours

Only one solvent is known to induce bladder tumours - diethylene glycol, a solvent used extensively in the food industry (7). The treated rats also had bladder stones. The role of foreign bodies in the causation of bladder tumours was investigated when it was shown that pellets of paraffin wax and cholesterol, which contained no added carcinogen, induced a low incidence of tumours when implanted surgically in the urinary bladders of rats and mice (9). This evidence suggested that the physical presence of a foreign body could induce tumours in the epithelium of the urinary bladder in rodents and there was therefore a strong suspicion that bladder stones could likewise induce such tumours. More convincing evidence was provided from an experiment in rats with ethylsulphonyl naphthalene sulphonamide. This compound when administered orally produced a high incidence of bladder tumours. It also produced an alkaline urine and bladder stones. Since the bladder stones were thought to be due to the alkaline urine, their formation was prevented by the administration of ammonium chloride to acidify the urine. Under these conditions no tumours were produced by the administration of the sulphonamide (8). This finding served to indicate that the bladder tumours which developed in rats given diethylene glycol almost certainly arose as a sequel to bladder stone formation. Low levels of exposure are therefore unlikely to present a carcinogenic risk to man.

Forestomach tumours

One solvent, 1,2-dichloroethane, is known to produce tumours of the forestomach in rodents when given orally. This anatomical area is unknown in primates including man and is thought to be the counterpart of the gizzard in ruminants. Irritants are known to produce a marked thickening of this region of the stomach and it is quite likely that prolonged irritation will lead to the appearance of papillomas and carcinomas. In any event it is difficult to see the relevance to man of tumours induced in an organ which does not exist in man. Although findings of this sort cannot be dismissed and should be supplemented by further appropriate studies to investigate the carcinogenicity of the compound in question, no judgement on the potential carcinogenic hazard to man is possible on the basis of the induction of this type of tumour in rodents.

Liver tumours

Most of the chlorinated hydrocarbon solvents induce hepatic tumours in laboratory rodents. The tumours induced are both benign and malignant. It is the usual practice to regard the induction of benign tumours, particularly in the mouse, to be of limited value in determining the carcinogenic activity of a chemical (6). Opinion is divided on the weight to be attached to the induction of malignant tumours. Some would regard them as incontrovertible evidence of carcinogenic activity with serious implications of a carcinogenic hazard to man. Others make a distinction between those compounds that induce highly malignant tumours of more than one type in the liver and those compounds that induce relatively well-differentiated tumours arising from hepatocytes only. This distinction is important because compounds which induce a diversity of tumours in the liver are generally also 'genotoxic', that is able to damage the genome (the site where genetic information resides within the cell), whereas the other class of compounds does not. The general tendency among oncologists is to consider that the genotoxic carcinogens carry a much more serious risk of being carcinogenic to man than the non-genotoxic ones. Genotoxicity is established if a number of mutagenicity tests are positive. These tests determine the ability of a compound to alter the hereditary characteristics of the test organism and are considered positive when one or more of these characteristics are altered. It is of interest that the chlorinated hydrocarbon solvents are not genotoxic and produce only tumours from hepatocytes which are either benign or well-differentiated malignant tumours. The action of non-genotoxic hepatocarcinogens in the rodent has not been accepted as being relevant to man in certain instances (29, 30).

Subcutaneous sarcoma

This tumour is produced by the injection of the test substance repeatedly at the same subcutaneous site in rats and mice. Carcinogens are known to produce this tumour but so do a number of other compounds without any suspicion of carcinogenic activity, such as glucose, hydrochloric acid and sodium chloride. Induction of this type of tumour is not regarded as a reliable index of carcinogenic activity (10).

CONCLUSIONS

The solvents that have produced tumours in animals have been in use for many years in industry, some of them before any real efforts were made to reduce human exposure so that one suspects that a fairly large number of workers have been exposed to fairly high concentrations for several years. Despite this, there have been no real indications, either from anecdotal or from epidemiological data, that they have induced cancer in man. This should not however lead one to adopt a complacent attitude towards these solvents but one should take appropriate advice to ensure that no unnecessary human exposure occurs.

One solvent which is associated with cancer in man is benzene. Exposure to this compound has been associated with the development of leukaemia in man (11). Fortunately the chemical is no longer as widely used as a solvent as it has been in the past. Interestingly this compound has not produced tumours in experimental animals. Although this exception does not invalidate the valuable contribution that animal studies could make in predicting a carcinogenic hazard to man it serves to emphasize that species may differ considerably in the carcinogenic response to chemicals. From this it would follow that interpretation of results from animal studies in terms of human hazard have to be carried out with great caution. Despite this, the experience gained over the last two or three decades has indicated that the carcinogens which are likely to present a real hazard to man possess certain characteristics, the principal ones being: (i) mutagenic activity in a number of systems which include mammalian cells or in vivo tests; (ii) carcinogenic activity in more than one species of laboratory animal; (iii) tumours are produced in more than one target organ; (iv) tumours exhibit a wide spectrum of severity - from highly undifferentiated to benign; (v) often more than one cell-type is affected in a target organ.

Attention to these points will serve to indicate whether exposure of workers to a carcinogenic hazard is likely to occur. Other studies will have to be carried out to ascertain this, such as pharmacokinetics at low and high levels of exposure, formation of adducts and detailed mutagenicity studies. Although this may appear to be a formidable programme there is at the moment no alternative for determining the presence of a carcinogenic hazard to man.

From the outline of the difficulties involved in interpreting animal studies presented in this paper it is fairly obvious that such results cannot be taken at face value. Equally, however, they cannot be ignored. The course of action outlined will go some way to providing information so that one can arrive at an informed judgement with regard to the possible carcinogenic hazard to man from exposure to solvents.

REFERENCES

1. Tomatis, L., Agthe, C., Bartsch, H., Huff, J., Montesano, R., Saracci, R., Walker, E. and Wilbourn, J. (1978). Cancer Res. $\underline{38}$, 877.
2. ECOTOC (1981). The Mutagenic and Carcinogenic Potential of Formaldehyde. Technical Report, ECOTOC, Brussels.
3. IARC (1974). Monographs on the evaluation of carcinogenic risk of chemicals to man. Vol.4. Lyon.

4. Young, S. and Hallowes, R.C. (1973). In "Pathology of Tumours in Laboratory Animals" Vol. 1. Tumours of the Rat. p.31. IARC, Lyon.
5. Grasso, P., Crampton, R.F. and Hooson, J. (1977). In "The Mouse and Carcinogenicity Testing". BIBRA, Surrey.
6. IARC (1979). Monographs on the evaluation of carcinogenic risk to man. Supplement 1, p.9.
7. Fitzhugh, O.G. and Nelson, A.A. (1946). J. ind. Hyg. Toxicol. $\underline{20}$, 40.
8. Flaks, A., Hamilton, J.M. and Clayson, D.B. (1973). J. nat. Cancer Inst. $\underline{51}$, 2007.
9. Bryan, G.T. and Springberg, P.D. (1966). Cancer Res. $\underline{26}$, 105.
10. Grasso, P. and Golberg, L. (1966). Fd Cosmet. Toxicol. $\underline{4}$, 297.
11. Infante, P.F., Wagner, J.K., Rinsky, R.A. and Young, R.J. (1977). Lancet $\underline{ii}$, 76.
12. Roe, F.J.C. and Grant, G.A. (1970). Int. J. Cancer $\underline{6}$, 133.
13. Campbell, J.S., Yang, Y.H. and D'Arcy Bolton, J. (1965). Arch. Path. $\underline{79}$, 500.
14. Kennaway, E.L. and Heiger, J. (1930). Br. med. J. $\underline{1}$, 1044.
15. NIOSH (1978). Current Intelligence Bulletin No. 25.
16. Morris, H.J., Nelson, A.A. and Calver, H.O. (1942). J. Pharmac. $\underline{74}$, 266.
17. Umeda, M. (1956). Gann $\underline{47}$, 153 & 595.
18. Prokofieva, O.G. (1962). Vop. Oncol. $\underline{8}$, 46 & 95.
19. Fujino, H., China, T. and Jami, T. (1965). J. nat. Cancer Inst. $\underline{35}$, 907.
20. Schmähl, D. (1964). Z. Krebsforsch. $\underline{66}$, 526.
21. Ponomarkov, V. and Tomatis, L. (1979). Scand. J. Work Environ. Hlth $\underline{4}$ Supp. 2, p.127.
22. Poel, W.E. (1963). Nat. Cancer Inst. Monograph $\underline{10}$, 611.
23. IARC (1976). Monographs on the evaluation of carcinogenic risk of chemicals to man. Vol 11. Lyon.
24. IARC (1972). Monographs on the evaluation of carcinogenic risk of chemicals to man. Vol 1. Lyon.
25. NIOSH (1978). Current Intelligence Bull. No. 20.
26. NIOSH (1978). Current Intelligence Bull. No. 25.
27. C.I.I.T. (1980). Report on Formaldehyde.
28. NIOSH (1978). Current Intelligence Bull. No. 30.
29. WHO (1979). DDT and its Derivatives. Environmental Health Criteria No. 9. UNEP/WHO, Geneva.
30. Cohen, A.J. and Grasso, P. (1981). Fd Cosmet. Toxicol. $\underline{19}$, 585.

PROTECTIVE ACTION OF 4-METHYL UMBELLIFERONE AGAINST THE INTOXIFICATION OF CARBON TETRACHLORIDE

A.Cerrati,P.A.Franco,M.Grasso,M.J. Keble, C.Puntrello,I.Raggi

Department of Toxicology, University of Milan, Italy.

1. INTRODUCTION

4-methyl umbelliferone (I) (Figure 1) is found in several aromatic plants of the Umbelliferous family (1). It has a pharmacological action on the biliary system (2). This property has been demonstrated in man by measuring changes induced in subjects undergoing surgical operations (3). Even at low doses (I) acts quickly to intensify and prolong biliary flow (5). Other workers have shown (I) to improve bilirubinaemia, prothrombinaemia and galactose clearance in treated subjects (6,7). It has an intense spasmotic action on the sphincter of Oddi(2). Some forms of hepatic damage seem to be reduced following treatment with (I). It appears to have a pronounced protective action on hepatic injury caused by bromobenzene in the rat (8). The severity

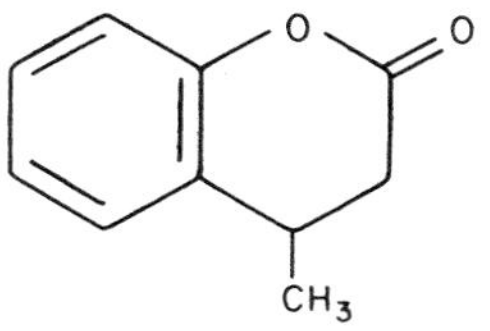

Fig 1, 4-methyl umbelliferone (I)

of fatty infiltration in the liver and of hepatic cell degeneration in animals treated with bromobenzene, decreases quickly in animals treated with (I). Rats treated with bromobenzene but protected with I show a normal parenchyma with a well preserved hepatic cellular structure without signs of fatty infiltrations and with a normal mitochondrial structure.

We have investigated whether (I) can protect against other substances which have a hepato-toxic action similar to that of bromobenzene.

The substance (I) has a highly effect protective action against some of the hepatic injury produced by administrations of ethanol to rats; it develops the onset of fatty infiltrations and other signs of hepatic injury (9). Since (I) protects animals against hepatic damage caused by ethanol it should also be effective against the liver injury caused by the administration of carbon tetrachloride. We have investigated this potential using the sodium salt of (I) which is particularly suitable for parenteral administration.

2. MATERIALS AND METHODS

Ninety male and female Sprague Dawley rats, average weight 200 g maintained on a standard laboratory diet, were divided into 3 groups each of 30 animals. Carbon tetrachloride was administered to the animals in the first group by gavage at a dose level of 1 ml of a 10% solution in olive oil/100 g body weight. The sodium salt of (I) in distilled water was administered to the animals in the second group by the subcutaneous route at a dose rate of 200 mg/kg/day for 4 days. On the fifth day carbon tetrachloride was administered to these animals by gavage at a dose rate of 1 ml of a 10% solution in olive oil/100 g bodyweight. The third group was used as untreated controls.

At the end of the study, 5 days after the administration of the carbon tetrachloride, all animals were sacrificed. The livers of the animals were subjected to gross and microscopic examination and a part of each liver analysed for lipid content using Parrings method.

3. RESULTS

The results show that subcutaneous administration of the sodium salt of (I) exerted a protective action against hepatic injury induced by carbon tetrachloride. In the animals dosed with carbon tetrachloride alone, hepatic damage was characterised by fatty infiltration, mitochondrial changes and parenchymal degeneration.

In animals treated with carbon tetrachloride but protected with (I) the parenchyma of the liver appeared unaffected with no evidence of fatty infiltration or damage to the mitochondria.

The total lipid content of the liver was increased more in the animals treated with carbon tetrachloride compared with those treated with carbon tetrachloride but protected with (I).

TABLE 1

Protective activity of 4-methyl umbelliferone (I) on hepatic injury caused by CCl_4 in rats.

No. animals	Treatment	Fatty Infiltration	Mitoch. alter.	Parenchimal damage
30	CCl_4	++++	++++	++++
30	CCl_4 +(I)	+---	+---	+---
30	---	----	----	----

TABLE 2

Lipid content in the liver of rats treated with CCl_4 and treated with 4-methyl umbelliferone (I).

No. Animals	Treatment	Hepatic Levels in mg/100 mg of weight
30	---	14 $\pm$ 1,6
30	CCl_4	51,2 $\pm$ 1,6
30	CCl_4 +(I)	22,5 $\pm$ 2,3

4. CONCLUSIONS

We conclude that 4-methyl umbelliferone used in our experimental scheme has been shown to have a protective action on hepatic cell injury caused by the administration of carbon tetrachloride. Carbon tetrachloride causes an increase in the levels of free radicals epoxides and fatty acids in the liver and also damages proteins, lipoproteins and subcellular structures. The observed parenchymal damage and microsomal alterations support this hypothesis.

It is believed that like vitamin E and ubiquinone, 4-methyl umbelliferone acts as an antioxidant and free radical trap.

REFERENCES

1. Cremoncini, C. and Toming, L. (1968). Minerva Medica 59, 4359.
2. Fontaine, L. et al. (1968). Therapie 23, 51.
3. Bonfils, S. and Maidensclaire, G. (1967). Therapie 22, 521.
4. Fontaine, L. et al. (1968). Therapie 23, 63.
5. Boissert, J.R. and Chivot, J.J. (1959). J. Physiol. (Paris) 51, 409.
6. Gaugand, J. (1969). L'Ovest Medical 22, 1483.
7. Tete, R. and Chatin, B. (1966). Cahier Med. Lyonnaise 13, 953.
8. Cerratti, A. et al. (1974). Proc. Euro. Soc. Study Drug Tox. 14, 237.
9. Cerratti, A. et al. (1975). Proc. Eur. Soc. Toxicol, 16, 213.

ABUSE OF SOLVENTS IN EUROPE

Christian Brule*

Council of Europe
BP 431 R6
F- 67.006 STRASBOURG CEDEX
France

1. INTRODUCTION

There has recently been, and in some quarters still is, what amounts to a moral panic about solvent sniffing. This panic is in some ways similar to the panic about illegal drugs in the 1960s, and in some ways different due to the changed circumstances of the 1970s. Since sniffing has occurred on and off in most urban areas for at least a decade, the sudden panic about it cannot be attributed to the existence of the activity itself. For reasons that I cannot discuss fully here, more attention is being paid to sniffing, and this attention is of a highly excitable, even "volatile" kind. A new crop of "glue experts" is being given media time, "glue terror" newspaper coverage increases, questions are asked in national parliaments and articles and pamphlets are written. But unfortunatley, often what is written on the harmful effects is unreliable and alarmist, tending to ignore distinctions between solvents and to atrribute to each and every one of them the combined total of possible ill-effects of them all.

2. HISTORY

It is many thousands of years since man first experimented with the naturally occurring substances in the environment in an attempt to relieve pain, overcome anxiety or alter his psychological state. In the acquisition of his experience he has encountered many substances which profoundly affect his nervous system. One method of introducing chemicals into the body is by inhalation of volatile substances. The large surface area of the lungs provides easy access and ensures a rapid onset of sensation. Many volatile substances

---

*The views expressed in this article are personal to the author.

have been exploited by man through the centuries and no age can boast an immunity from their abuse. During the present century the deliberate misuse of volatile agents by adults has not disappeared entirely but seems to have occurred only sporadically. The earliest substances to be abused were the anaesthetics. In the early 1800s the first uses of the anaesthetics nitrous oxide, ether, and chloroform were as intoxicants. Medical applications in the fields of surgery and dentistry followed. The use of anaesthetics for recreational purposes, laughing gas, continued throughout the 19th century in Europe and America and even exhibited an occasional resurgence in this century in the 1920s and 1940s. Abuse of volatile hydrocarbons is not new and ether abuse was common in Europe and North American in the 1800s. Widespread sniffing of plastic model glues and nail polish removers began in the 1960s.

3. DEFINITION OF SUBSTANCES

Volatile substances are chemicals that vaporise to a gaseous form at normal room temperatures. When inhaled they can intoxicate. While most of the substances currently abused are commercial preparations that are generally safe when used as directed for their intended purpose, and often do not result in permanent injury even when they are abused, certain such as the aerosols have proved to be harmful or even fatal.

Abused volatile substances can be classified under three headings: anaesthetics, solvents and aerosols. Volatile hydrocarbons are organic chemicals produced from petroleum and natural gas. Because they are volatile and evaporate quickly at room temperature, they are popular in the marketplace as a base for fast-drying products. Solvents are used in both industrial and household preparations. They include plastic cement (hexane), model airplane glue and lacquer, thinners (toluene, xylene), nail polish remover (acetone), lighter fluid (naphtha), cleaning fluid (benzene, trichloroethane) and gasoline. An aerosol is defined as a liquid, solid, or gaseous product discharged from a disposable container by a compressed gas propellant. Cookware coating agents, deodorants, hair sprays, insecticides, medications, and paint are just a few examples of aerosol products. Since the widespread development of aerosol sprays, abuse of them has also become a problem. It primarily involves sniffing fluorocarbons, the gases used to propel a small number of aerosol products. These gases have been associated with fatal sniffing accidents.

SOLVENT ABUSE

The most common volatile substances of abuse are: thinner, rubber, cement, glue, trichloroethylene and gasoline. Many other substances of abuse exist but reference will only be made to those most commonly used.

Trichloroethylene and Other Chlorinated Hydrocarbons

Trichloroethylene is sold in shops as a spot remover and is found in glues and cements. It is used professionally in car workshops, in tyre shops, by sprayers, tailors, clothing manufacturers, dry-cleaners and as an anaesthetic by doctors. One of the biggest advantages of tri is its ability to dissolve fats; it is more or less non-ignitable. Trichloroethylene is a chlorinated solvent, a group which even includes methyl chloroform, chloroform, perchloroethylene and carbon tetrachloride.

Trichloroethylene, methyl chloroform and perchloroethylene have been the cause of a number of sudden sniffing deaths, i.e. rapid death, owing to respiratory depression or cardiac arrest, in immediate conjunction with sniffing, visual disturbances, pulmonary oedema and damage to the liver, kidneys, brain and myocardium may also ensue. The risks are potentiated by alcohol consumption. Methyl chloroform is often found in aerosol spot removers in which the combined effect of propellant gas and methyl chloroform increase the risk of SSD.

Toluene

Toluene is on retail sale mainly in paint thinner and other substances for the dilution of paint and cleaning of brushes; it is also found as a solvent in paints and some glues, often as a substitute for trichloroethylene. It is also used professionally as a solvent for paints, rubber, resins and oils and in the cleaning of machines. The main risk with toluene is presented when the toluene is contaminated with benzene. The deaths reported in which toluene was involved have entailed cases of advanced mixed abuse of narcotics, alcohol and thinner or have been accidents or suicides in which the victim had drunk thinner.

Xylene

Xylene is used for about the same purposes as toluene and presents the same risks.

Benzene

Benzene is classified as a poison. Damage to haematogenic organs with pernicious anaemia, liver and kidney damage have been reported.

Petrol

Petrol usually contains varying amounts of aromatic hydrocarbons such as benzene, toluene, xylene, tetraethyl lead, compounds of bromine and chlorine. The damage (brain damage) caused by the lead compound must be added to the risks described above for benzene.

The benzene concentration in petrol depends on the composition of the crude and the duration of the refining process. Refining promotes the formation of benzene. So shorter refining processes yield less benzene but a poorer quality petrol (lower octane) calling for the use of more lead additive. Purifying petrol from benzene is a very expensive process. It might be possible to reduce the benzene concentration in petrol by employing other refining methods.

Petrol Sniffing - the first cases described were concerned with the inhalation of gasoline fumes in America during the 1950s. Despite the fact that petrol or gasoline was then and still is readily available at little or no cost, less than 30 cases had been recorded since the practice was initially reported.

The literature available offers information gleaned from case reports on small numbers of sniffers referred for psychiatric assessment and treatment on other grounds, for example strange behavioural patterns or hallucinations, most visual but occasionally auditory. Gasoline or petrol sniffing is predominantly a male activity and most of the reported cases occurred in urban areas. The age range of cases at the time of referral has varied between 6 and 20 years.

The method employed usually involves direct inhalation from the petrol tanks of cars or motor cycles. Although exact information is frequently omitted by authors, the degree of involvement in the practice has ranged from one every few months to three or four times weekly. While the term addiction has been applied by some authors to chronic sniffers of gasoline, no withdrawal symptoms occurred when the patients stopped the practice. The home environment of many gasoline sniffers would appear

SOLVENT ABUSE

to have been unstable and often characterised by parental absence, family discord and alcoholism in one or other or both parents.

In view of the fact that gasoline or petrol is a mixture of saturated and unsaturated hydrocarbons with tetraethyl lead and tricresyl phosphate added, it is perhaps surprising that toxic systematic damage resulting from gasoline sniffing occurs infrequently. The absence of such damage may be partially explained by the very crude method of inhalation and the fact that the practice was intermittent and carried on in the open air. In one of the cases reported, the sniffer developed peripheral neuritis, thought to be due to the tricresyl phosphate component of petrol and in a further two cases there was clinical evidence of a chronic brain syndrome, thought to be caused by the toxic effects of eptrol. In addition, four teenagers were reported as having suffered severe burn injury as a result of gasoline sniffing, thus demonstrating very clearly the dangers of misusing a flammable substance.

Sniffing of Glue and Similar Substances

In 1959 glue sniffing or, more specifically, the inhalation of vapours from glues or plastic cements made its appearance. The first cases were reported in California but the practice seems to have spread to the midwest in the early 1960s. By 1965 it was reported to be occurring in every state of the United States of America. By the end of the 1960s, it had involved children and adolescents in countries as far as scattered as Africa, Australia, Canada, Finland, Japan, Mexico, South America and Western Europe.

Besides glues and cements a wide variety of other substances have been exploited. These include nail polish removers, antifreeze, trichloroethylene, chloroform, paint-thinner, lacquer-thinner, marking pencils, cleaning fluids, amyl nitrite.

Solvent

The organic solvent most often found in adhesives and plastic cements is toluene, though of course there are others including benzene, cyclohexane, hexane, tricresyl phosphate and xylene in various combinations. These are also likely to be present in lacquer thinners and enamels. Cleaning fluids, on the other hand usually contain at least one from the following - perchlorethylene, trichloroethane, and carbon tetrachloride. Lighter fluids contain a mixture of aliphatic hydrocarbons particularly naphtha, while nail polish removers contain acetone and aliphatic acetates. Freons, a series of fluorinated-chlorinated hydrocarbons, are used to propel aerosel contents and may also be used as refrigerants.

Characteristically these solvents are readily soluble in fats, poorly in water and have low boiling points which make them volatile at room temperatures. They affect the lipid components of cells, especially those of the central nervous system.

4. EFFECTS

The Direct Effects of Abuse

The effects of any drug depend on the amount taken at one time, the past drug experience of the user, the circumstances in which the drug is taken (the place, the feelings and activities of the user, the presence of other people, the simultaneous use of alcohol or other drugs etc.) and the manner in which the drug is taken. Inhaled vapours from solvents or aerosols enter the bloodstream rapidly from the lungs and are then distributed most rapidly to organs with a large blood circulation such as the brain and the liver. Most volatile hydrocarbons contained in solvents and aerosols are fat soluble and thus abosrbed quickly into the central nervous system, producing depression of many body functions, including respiration and heart beat. Accumulation in fatty tissue occurs less rapidly.

While some volatile hydrocarbons are metabolised and then excreted through the kidneys, many are eliminated unchanged, primarily through the lungs. Because of this, the odour of solvents will remain on the breath for several hours following inhalation. The complete elimination of volatile hydrocarbons may take some time, as redistribution proceeds slowly from fatty tissues.

Short-term effects are those which appear rapidly following inhalation and disappear within a few hours or days. The initial effect of inhalation is a feeling of euphoria characterised by lightheadedness, pleasant exhilaration, vivid fantasies, and excitation. Nausea, sneezing and coughing, hallucinations, increased salivation and sensitivity to light may also occur. In some individuals, feelings of recklessness and invincibility may lead to bizarre behaviour.

SOLVENT ABUSE

Deep inhalation or sniffing repeatedly over a short period of time may result in disorientation and loss of self-control, unconsciousness, or seizures. Muscular incoordination and depressed reflexes are characteristic of this stage. Nosebleeds, bloodshot eyes, unpleasant breath, and sores on the nose and mouth may also occur.

The effects of an initial, brief inhalation fade after several minutes, but concentrating the drug inside a plastic bag, for example, may prolong the effects for several hours. An experienced user can maintain a "high" for as long as 12 hours by periodic sniffing. Brain depression leading to unconsciousness rarely occurs. For the majority of users, most effects pass within an hour after sniffing is discontinued. Hangovers and headaches lasting several days may follow use, although less commonly than after alcohol consumption.

Sniffing of solvents and aerosols has been associated with many fatalities. The most common type is the "sudden sniffing death" which occurs most frequently during abuse of aerosol sprays (fluorcarbons), spot removers (trichlorethane, carbon tetrachloride), and model airplane cement (toluene, acetone). It is believed these substances cause the heart to react abnormally, especially to stress or intense exercise, causing irregular heart beat (arrhythmia) which may result in sudden death. Asphyxia or suffocation due to sniffing solvents from a plastic bag may also be fatal. Some accidental deaths have resulted from bizarre behaviour caused by sniffing.

Long-term effects are those which appear following repeated use over a long period of time. These include pallor, fatigue, forgetfulness, inability to think clearly or logically, tremors, thirst, weight loss, depression irritability, hositility, feelings of persecution, and reduction in the formation of blood cells in the bone marrow caused by aromatic hydrocarbons such as benzene. Many of these long-term effects are reversible if drug use is stopped.

Specific Effects

Toluene - in the few cases of acute toluene poisoning reported (industrially) the effect has been that of a narcotic, the workman passing through a state of intoxication into one of come. Recovery following removal from exposure has been the rule.

Exposure to concentrations up to 200 ppm produces few symptoms. At 200 to 500 ppm, headache, nausea, loss of appetite, a bad taste, lassitude, impairment of coordination and reaction time are reported, but are not usually accompanied by any laboratory of physical findings of significance. With higher concentrations, the above complaints are increased and in addition anaemia, leucopenia and enlarged liver may be found in rare cases".

Therefore, liver damage, impairment of production of red blood cells from bone marrow, and anaemia, are possible but only with repeated use. Frequent glue sniffers develop a tolerance to the effects of toluene, and need a greater amount of the product each time if it is used frequently. Thus, there is an eventual increase in the ppm rate (parts per million of glue in relation to air).

In paint containing lead, and in lead-containing gasoline (even low-lead), the lead itself overshadows the petroleum base in terms of relative danger. Lead is a cumulative property and is not eliminated from the body so lead intake can be lethal.

Non-lead paints and petroleum products are a bit more complex, but nearly as damaging in the long run. Sniffing of acrylic spray lacquer is becoming common in some areas, and in certain cultural-economic circles. Virtually all paints and petroleum distillates have an overdose potential. The only exception is latex paints, which will not produce states of intoxication because it is not a petroleum product.

Problems of Tolerance

Gasoline, paint, petroleum products, glue and other substances contain volatile hydrocarbon solvents which are highly soluble in lipids (fats) - a major component of living tissue. Nearly all of these solvents are stored in the body long enough to create a tolerance with frequent use, and so toxicity gradually goes up to danger levels. However, though tolerance is the cause, there is yet no proof of withdrawal symptoms, and the term "physical addiction" does not appear to apply.

Dependence

Regular use of inhalants induces tolerance, making increased doses necessary to produce the same effects. After one year a glue sniffer may be using many tubes of plastic cement to get the "high" originally obtained with a single tube.

Inhalants and Pregnancy

Little is known about the effects of inhaled volatile hydrocarbons on pregnancy and foetal growth.

5. EXTENT OF ABUSE

Who Uses Inhalants?

The majority of users range in age from 8 to 16, with an average of 12 to 13. However, some heavy users are in their late teens, early twenties, or even older. A Canadian 1979 Addiction Research Foundation survey of drug use among Ontario students in grades 7 to 13 showed 4.3% had used glue and 6.2% had used other solvents at least once during the preceding year. Among those age 12 and 13, the rate of use was highest at 7.1% for glue and 9.3% for other solvents.

Because solvents are widely available and easy to obtain, some users are found in prisons, factories where solvents are manufactured and processed, and Indian reserves where gasoline sniffing has sometimes reached epidemic proportions. Most users come from families where one or both parents are absent. Histility and lack of affection are common characteristics of their background, although not specific to them. More boys than girls inhale solvents.

Why do People Use Inhalents?

People report various reasons for abusing these substances. Curiosity and social pressure are factors, although many stop once their curiosity is satisfied. Those who continue drug use report that they like the "high" following inhalation. Others claim to use solvents to reduce anxiety or depression, to compensate for feelings of inferiority, shyness, insecurity, or to relieve boredom.

Solvent abuse has been associated with antisocial behaviour such as dangerous driving, larceny, property damage, shoplifting, and theft. It is not clear that the drugs cause such behaviour. More likely, sniffing is more common among those who would engage in these acts in any case.

Those who sniff give many reasons for their activity such as:

"everyone else was doing it"
"it just made me forget the reason I was made, I took it to cool off"

"there wasn't anything else to do"
"if I didn't do it they called me chicken"
"just for kicks, just to be with the crowd"
"I'm not old enough to drink beer so I sniff, I just did it when I was lonely or bored or mad, if you sniff, you forget about your problems"
"I don't know"
"It's easier to buy than anything else"

In many cases, adolescents who sniff solvents come from homes in which one parent, often the father, is absent or alcoholic, where there is frequent family strife and where there are often chronic social and economic problems. For reasons that we do not at present understand such conditions particularly predispose adolescents to psychological dependence on solvents. Sniffing is often the result of peer or group pressure. Depending on one's emotional balance sniffing may simply be a short-term adventure or develop into a longer-term problem.

6. DETECTION

By far the best way to detect solvent abuse is to have a reasonable level of suspicion and to ask. Most persons who are engaged in sniffing will readily admit to it when asked, though they may not spontaneously approach anyone with a request for help if left to their own devices.

Evidence of solvent sniffing may be found in the smell of solvent vapours on the breath or about the clothing. One may also find solvent soaked rags, paper or plastic bags and empty containers of solvent and/or aerosol products lying about.

The sniffer may exhibit the following physical symptoms (each of which may be due to other causes):

excessive running of nose and eyes
persistent coughing
upset stomach
poor appetite, weight loss, thirst
irritability and inattention
muscular incoordination
drowsiness
unconsciousness

In addition, school work or job performance may have been deteriorating for some time. Associated with this complex of symptoms, the sniffer may demonstrate particular antisocial delinquent behaviour. Tossing glass bottles, intimidating neighbourhood residents, creating noisy disturbances and shoplifting are just a few examples. Interestingly enough the sniffer may apologise for this behaviour a day later and repeat the same activity thereafter.

7. PREVENTION AND CONTROL

What Kinds of Programme can be used in Raising Community Awareness to Reduce solvent and Aerosol abuse?

It is worthwhile to consider that sniffers tend in general to be young, ranging in age from 8-16, although some are in their early twenties. Males tend to predominate. Because the sniffing environment may vary from isolated rural to larger urban centres, programmes aimed at helping should be selectively tailored to the area and group involved. We do not want to tune-in kids who are otherwise tuned-out.

Experience demonstrates that it is advisable to employ a low-key local strategy in combatting solvent or aerosol abuse. In particular, community approaches should be emphasised and implemented where possible (see Acknowledgements).

Solvent or aerosol products in the household should be removed to a secure place where access is restricted. Parents should request their local retailers to exercise caution and restraint in light of solvent and aerosol abuse. Furthermore, manufacturers of these products should be informed of sniffing incidents so that they can maintain an awareness of such abuse and consider this factor in their overall marketing strategy. Voluntary co-operation and statutory controls among manufacturers and retailers to limit the access of potential abuses to these products appears necessary and has borne fruit in several countries. An example is the denaturing of sniffable products, i.e. by the addition of some unpleasant smelling or irritating substance in order to make sniffing impossible. There are several examples or denaturing practised; glues have been denatured with mustard oil in the USA, and an aldehyde has been used in thinner in Norway. Ammonia has also been tried as denaturing agent.

8. CONCLUSIONS

Solvent and aerosol abuse have proved to be difficult and refractory problems. What is needed is the constructive application of creative measures developed in concert by many different elements of the community. When such approaches are developed, some thought should also be given to involving persons with expertise in evaluative research to determine whether or not these approaches are in fact effective. The need for the development of effective approaches in this area is urgent, and careful evaluation will be a key to future progress. However, sensationalising the matter should be avoided. Attempts at politicising the issue should likewise be avoided.

9. ACKNOWLEDGEMENTS

We want to pay special tribute to:

- The Addiction Research Foundation in Toronto for the two brochures "Facts about inhalants (solvents and aerosols" - revised January 1980; and "Information review on solvent and aerosol abuse"- February 1977.

- Dr. Watson J.M. (University Dept. of General Practice in Glasgow) for his article on "Solvent abuse by children and young adults" (British Journal of Addiction, Vol. 75 No. 1, March 1980)

parts of which have been used for this article.

SNIFFING 1,1,1 - TRICHLOROETHANE
SIMULATION OF TWO FATAL CASES

P.O. Droz, C. Nicole, E. Guberan*

*Institute of Occupational Medicine and Industrial
Hygiene of the University of Lausanne, Switzerland*

* *Occupationl Health Service of
the Canton of Geneva, Switzerland*

1. INTRODUCTION

The sudden sniffing death syndrome (SSD) was brought to light by Bass (1) as a result of his analysis of 110 fatal cases which occurred during the 1960's to American youths. The usual sequence of events was : a deliberate inhalation of an organic solvent, a short period of emotional or physical stress and a sudden death without any anatomical cause at autopsy. Similarly to anesthesic death, SSD was attributed to the sensitization of the heart to catecholamines-induced arrythmias by volatile hydrocarbons. Among the 110 fatalities, 29 were related to 1,1,1-trichloroethane (TCE) alone, 2 to TCE and toluene and 59 to aerosols in which TCE is frequently added to the pressurised fluorocarbons serving as propelling agents. Thus, in this series TCE was the compound most frequently associated with SSD, although it doesn't seem that sniffers have used TCE more often than other solvents (2,3).

Despite the numerous cases already described in the literature concerning TCE or other solvents, it is known in one report only (toluene sniffing) (2) what the inhaled concentration was during the fatal episode.

In recent years, we have investigated in Geneva two cases of SSD, reported below. Both were due to TCE and these are the only sniffing deaths known in Switzerland where solvent sniffing seems to have occurred only sporadically. The aim of this study is to estimate the level reached by TCE before death for these two cases. We therefore have reproduced experimentally the conditions under which TCE inhalation occurred.

2. DESCRIPTION OF CASES

Case I

During lunch break, a 17-year old unskilled worker filled a six-litre can with two litres of TCE, put it on the work-bench of a small repair-shop, locked the door and inhaled when seated. After a time that could not be determined (between 5 and 45 minutes), his foreman caught sight of him through the door-pane, rushed to the door and knocked on it. The young worker had his head on the can, was crying and his shoulders were shaking with spasms. The foreman ran back and entered the room through the window. A first helper was called who, finding neither respiration nor pulse started mouth to nose respiration. The coronary care ambulance men arrived 15 minutes later. They started full reanimation and obtained a spontaneous cardiac activity. At the hospital, however, cerebral death was diagnosed. No anatomical cause of death was found at autopsy.

Case II

A 20-year old apprentice mechanic inhaled a rag soaked with TCE. He vomited and collapsed about five minutes later while speaking with his two fellow workers. One of his fellows started mouth to nose respiration and cardiac massage was carried out by the ambulance men. On arrival at the hospital half an hour after the collapse, a ventricular fibrillation was recorded without any sign of life. In spite of eight cardiac electric shocks, the reanimation failed. No anatomical cause of death was revealed by the autopsy.

This case has been published in greater detail elsewhere (4).

3. EXPERIMENTAL SIMULATION

General strategy

In order to take into account any perturbating factor, standard conditions have been decided for both cases.
All the simulations have been carried out in an experi-

mental chamber (5) of 10 m³ with an air exchange rate of 10 volumes per hour. Drafts were minimized so that they would not affect the concentration of solvent.

The sniffer himself was simulated using a dummy head. A hole was drilled between the mouth and the nose to include a piece of tubing for simulation of breathing. Air was sampled in this tube at a flow rate of 200 ml/min for analysis.

A Perkin-Elmer 900 gas chromatograph was used, equiped with 2 FID detectors, and gas sampling valves. The transfer teflon lines and the valves themselves were heated at about 50°C to avoid adsorption. Calibration was done using standard mixtures in air prepared beforehand by injecting various amounts of TCE into a 50 l. flask.

The respiration was simulated using a hand operated bellow type pump of variable volume. The volume and flow profiles were checked with a Fleish spirometer.

The overall reproducibility of the experimental system was tested on several occasions.

Case I protocol

A can (⌀ 20 cm, 6 l.), the same as that employed during the fatal case, was used for the simulation. Two litres of technical grade TCE were added into it. The temperature of the liquid was monitored periodically and remained between 20 and 21° C.

As the dummy head can act as a lid on the can, and therefore has a large influence on the concentrations inhaled, 3 different head positions were investigated, i.e. 0, 5 and 10 cm from the opening of the can.

Several breathing patterns have been tried in order to detect any significant influence : respiratory frequency 0 (no breathing), 12 and 24 min^{-1}, tidal volume 600 and 1500 ml.

In order to identify any eventual trend in the concentration, each simulation was carried out for a period of about 30 minutes representing 10 to 20 GC analyses.

Case II protocol

The rag used was cotton, about 500 cm^2 in size. It was impregnated by dipping into TCE and was then squeezed to remove excess solvent. The reproducibility of this technique was tested by weighing the impregnated rag.

It was then fixed immediately onto the mouth-nose region of the head. Several breathing patterns were again tested : respiratory frequency 12 and 24 min^{-1}, tidal volume 600, 1200 and 1500 ml. Furthermore inspiration and expiration as well as inspiration only through the rag were tested.

About 10 GC analyses were carried out in each simulation during a period of between 20 and 40 minutes.

4. RESULTS AND DISCUSSION

The different simulations carried out for case I are presented in table 1 together with the mean concentrations and the standard deviations obtained. The situations with no respiratory frequency and tidal volume correspond to the no breathing case. This can give an idea of the influence of pulmonary ventilation on the levels of solvent. Fig. 1 and 2 present the individual results in greater detail.

Table 1

Results of case I

No	Distance (cm)	Respiration simulation (min^{-1})	(ml)	Mean TCE (ppm)	Standard deviation (ppm)
I	0	-	-	40500	1900
II	0	-	-	34600	3800
III	0	12	600	34400	2690
IV	0	24	600	24500	1900
V	5	-	-	3990	1590
VI	5	12	600	9500	2600
VII	5	24	600	6460	760
VIII	5	12	1500	8930	2300
IX	10	-	-	10300	6800
X	10	12	600	2850	1200
XI	10	24	600	3230	570

Despite the wide variations of concentration, it can be seen that the levels are quite stable during the experiments and that they are reproducible (table 1, situations I and II). In both figures 1 and 2, the points separate clearly into three sets corresponding to the 3 distances between the can and the dummy head. This parameter has therefore a prevailing impact on the inspired concentrations. On the other hand, in Fig. 2, the breathing pattern does not seem to have a large influence. The comparison of Fig. 1 and 2 indicates that breathing becomes an important parameter only when the head moves away from the can.

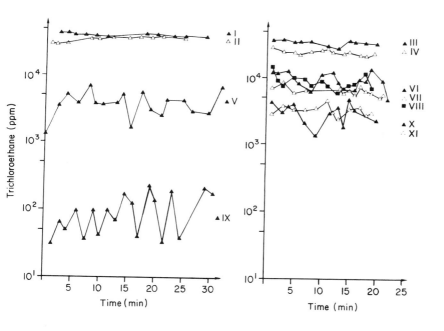

Fig. 1 Case I, no breathing   Fig. 2 Case I, breathing

The experiments relating to case II are summarized in table 2 and the results obtained are presented in Fig. 3 and 4. The reproducibility of the rag impregnation was of the order of ± 5% as measured by weighing. Contrary to what was observed in case I, there appears to be a definite trend in the levels of TCE corresponding to the progressive evaporation of the solvent from the cloth. The comparison of the different situations does not show any significant influence of the parameters tested such as type of breathing, respiratory frequency and tidal volume.

Table 2

Simulations of case II

Situation No.	Type of breathing	Respiratory frequency	Tidal volume
I	insp./exp.	12	1500
II	insp./exp.	12	1500
III	insp./exp.	12	600
IV	insp.	24	600
V	insp.	12	600
VI	insp.	12	1200

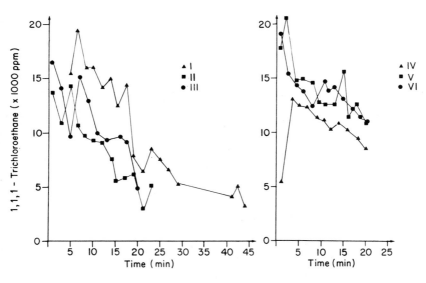

Fig. 3 Case II, insp.+ exp.   *Fig. 4 Case II, insp.*

As can be seen from the results obtained in both cases I and II, the range of concentrations is very wide. Nevertheless, assuming that in case I, the probable distance between the can and the head during the fatal sniffing was of the order of 5 cm, a concentration between 6'000 and 14'000 ppm can be expected (Fig. 2). Similarly in case II it is reasonable to suppose that sniffing time was no longer than 10 minutes, giving a concentration varying between 10'000 and 20'000 ppm. Therefore, despite the large differences in the sniffing procedures for these two cases, the levels of TCE observed are quite similar.

The range of concentration found agrees well with the concentration of 10'000 ppm reported by Press and Done (2) for toluene after putting a gauze soaked with polystyrene cement in a paper bag, as often practised by sniffers.

Animal experiment has shown that the minimal inspired concentration necessary to produce epinephrine-induced arrythmias in the dog was 5'000 ppm of TCE (6). Our data (6'000 to 20'000 ppm) seems to indicate that similar concentration levels are responsible for SSD in human.

5. REFERENCES

1. Bass M. (1970). *JAMA* 212, 2075-2079
2. Press E., Done A.K. (1967). *Pediatrics* 39, 611-622
3. Watson J.M. (1980). *British Journal of Addiction* 75, 27-36
4. Gubéran E., Fryc O., Robert M. (1976). *Schweiz. med. Wschr.* 106, 119-121
5. Guillemin M. (1975). *Arch. Mal. Prof.* 7-8, 421-428
6. Reinhardt C.F., Mullin L.S., Marfield M.E. (1973). *J. Occup. Med.* 15, 953-955.

THE UNITED STATES APPROACH TO THE SETTING OF STANDARDS

Marcus M. Key, M.D., M.I.H.

*University of Texas Health Science Center at Houston
School of Public Health
P.O. Box 20186
Houston, Texas, U.S.A.*

In the United States, we have two approaches to the setting of standards: the official approach under OSHA, our acronym for both the Occupational Safety and Health Act of 1970 and the Occupational Safety and Health Administration in the Department of Labor, and the voluntary approach by such organizations as the American National Standards Institute (ANSI) and the American Conference of Governmental Industrial Hygienists (ACGIH). The Occupational Safety and Health Act mandated start-up standards and permitted recognized Federal standards and national consensus standards to be incorporated. Accordingly, some 360 Federal health standards promulgated previously under the Walsh-Healey Public Contracts Act were incorporated along with 22 health standards from ANSI as start-up standards under OSHA. The former were the 1968 Threshold Limit Values (TLVs) of ACGIH, and the latter were Maximum Allowable Concentrations (MACs) which also contained time-weighted average limits. The old ACGIH TLVs and ANSI MACs are still being enforced by OSHA. Some 20 new standards have been promulgated by OSHA, practically all of them carcinogens. There was also a new standard for benzene, but it was remanded to OSHA in a 1980 Supreme Court Decision because the new standard of 1 p.p.m. was not sufficiently supported by scientific evidence.

Under a provision of the Occupational Safety and Health Act, the States are permitted to develop and enforce their own occupational safety and health standards provided they are at least as effective in providing safe and healthful employment and places of employment as the standards promulgated under Federal OSHA. Almost half of the States and other jurisdictions included under the Act operate enforcement programs under matching grants from Federal OSHA. Most

State programs have adopted the Federal OSHA health standards by reference, but a few States do not routinely adopt Federal standards.[1] One State in particular, California, has a reputation for being somewhat tougher than Federal OSHA, and this is illustrated in Table 1, which compares occupational exposure limits for solvents among the top 50 chemicals as listed by the Chemical & Engineering News.[2] Most large companies that have plants in several States prefer to have standard setting by Federal OSHA and enforcement by State OSHAs, whereas organized labor prefers Federal standards and enforcement.

The ACGIH TLVs are still the most up-to-date permissible exposure limits available in the U.S., and most large corporations with industrial hygiene programs adhere to the stricter ACGIH TLVs because they represent the latest word on good industrial hygiene practices. The Chemical Substances TLV Committee has a Subcommittee on Hydrocarbons and Oxygenated Hydrocarbons, chaired by Prof. James W. Hammond of the University of Texas School of Public Health.[3] This Subcommittee meets twice a year separate from the full committee and does an excellent job of keeping up with the latest toxicologic and epidemiologic studies on solvents. ACGIH TLVs are reviewed annually, but the OSHA standard setting process is cumbersome and has become very difficult in recent years because of polarization over the issue of potential carcinogens, the adversarial approach to occupational safety and health in the U.S., and the increasing tendency to challenge standards in the courts. Figure 1 shows the intended sequence of steps for standard setting under the Act compared with the present situation, which reflects the realities of our adversarial approach and litigous society. It should be noted that the Secretary of Labor can consider other inputs, such as information submitted in writing by an interested person, a representative of employers or employees, a nationally recognized standards-producing organization, etc. In recent years, OSHA has been reluctant to appoint advisory committees and has largely ignored the NIOSH Criteria Documents with their recommendations for standards. The Reagan Administration requires development of Regulatory Impact Analyses (RIAs) both before publishing proposed rules and before adopting final rules. These RIAs and the rules to which they pertain must be forwarded to the Office of Management and Budget for regulatory review and suggestions as to how they might be improved. These are intended to ensure that the prospective benefits of each rule shall outweigh its

TABLE 1

SOLVENTS AMONG TOP 50 CHEMICALS (C&EN, 1981)

1980 Rank	Solvent	Permissible Exposure Limits in PPM*		
		OSHA	CalOSHA	ACGIH TLV
15	Toluene	200	100	100
16	Benzene	10	10	10
17	Ethylene dichloride	50	50	10
18	Ethyl benzene	100	100	100
23	Xylene	100	100	100
28	Ethylene glycol	—	100	50
38	Acetone	1000	1000	1000
39	Cyclohexane	300	300	300
41	Vinyl acetate	—	10	10
43	Isopropyl alcohol	400	400	400
49	Ethanol	1000	1000	1000

*Time-weighted average

Intended	Current
NIOSH-OSHA Priority Assessment ↓	
NIOSH Criteria Document ↓	
OSHA Advisory Committee ↓	Occupational Disease Outbreak ↓
Proposed OSHA Standard ↓	Proposed OSHA Standard (RIA)* ↓
OSHA Public Hearing ↓	OSHA Public Hearing ↓
Final OSHA Standard	Final OSHA Standard (RIA)* ↓
	Circuit Court of Appeals ↓
	Supreme Court

* Regulatory Impact Analysis

Fig. 1. U. S. Approach to Standard Setting Under OSHA

THE US APPROACH TO STANDARD SETTING

prospective cost, and that each rule shall represent the regulatory alternative least burdensome to society.[4]

Very few occupational health standards are expected from OSHA in the next few years. Occupational health activities within the Federal government appear to be entering a period of (1) reduced resources, (2) regulatory de-emphasis, (3) programmatic redirection, and (4) re-interpretation of the Federal role in controlling workplace hazards.

According to a recent OSHA policy decision, all new standards must meet the following requirements:
(1) OSHA must demonstrate a significant risk.
(2) OSHA will analyze scientific data objectively and examine economic and technological feasibility on an industry-wide basis to set a level of exposure to protect employees.
(3) OSHA must demonstrate that the proposed standard will reduce the risk.
(4) OSHA will analyze the cost-effectiveness of proposals to determine the best way to achieve the level of protection.[5]

A logical approach to the setting of standards requires a priority system based on relevent and defensible parameters. OSHA apparently has had no priority system for setting health standards, preferring instead to react to pressures from crises and the courts. NIOSH's original priority system for developing Criteria Documents and programming supportive research was based on the five indices of population exposed, relative toxicity, incidence, quantity, and trend. Lacking some of the information needed for these parameters, the initial priority list was determined by multiplying the estimated number of workers exposed by a severity rating, which was assigned by a panel of experts from within and outside the government. The NIOSH priority system was refined and updated as new data became available from NIOSH's first National Occupational Hazard Survey (NOHS I) and the National Cancer Institute's Carcinogenesis Bioassay Program. A second National Occupational Hazard Survey, now called the National Occupational Exposure Survey (NOES), begun in 1980, should provide even better data on potential occupational exposures. From NOHS I data it can be estimated that some 3.5 million workers were <u>potentially</u> exposed, full or part-time, to solvents in the early 1970's. The base population for this study was 38 million, so the percentage of the U.S. workforce with <u>potential</u> exposure to solvents was 9.1%. I should emphasize that the NOHS I and NOES are only valid for potential exposure--the surveyors did not distinguish between controlled and uncontrolled exposures.

NIOSH is developing a new system of priority setting and quantitative risk assessment which will incorporate all NIOSH surveillance activities and include general industry health and safety risks, as well as mining risks. There will be five lists, and one of these will be an overall master list of priority subjects for criteria or other document production. A distinction will be made between quantitative risk assessment, which is NIOSH's responsibility, and risk evaluation, which is OSHA's responsibility.[6]

Two concepts have recently been proposed that are applicable to the establishment of priorities for solvents. One is the Risk Ratio developed by Professor Hammond to aid in selection of solvents for safe industrial use:

$$\text{Risk Ratio (RR)} = \frac{\text{Evaporation Rate (ER), referenced to Butyl Acetate as 100}}{\text{Threshold Limit Value (TLV)}}$$

Risk Ratios for several kinds of solvents are shown in Table 2.[7] These were originally based on 1979 TLVs, and some would have to be recalculated.

The other concept that can be used to assess priorities was developed for animal carcinogens by Dr. Robert A. Squire, Associate Professor of Comparative Medicine at Johns Hopkins School of Medicine and former Acting Director of the Carcinogenesis Testing Program of the National Cancer Institute.[8] This method of ranking animal carcinogens was developed to counter the assertion that all animal carcinogens pose equal threats to human health, regardless of variations in the evidence. Squire's method assigns scores, or numerical ratings, according to responses of the animals to the following six criteria: (1) number of different species affected; (2) number of histogenetically different types of neoplasms in one or more species; (3) spontaneous incidence in appropriate control groups of neoplasms induced in exposed groups; (4) dose-response relationships (cumulative oral dose equivalents per kilogram of body weight per day for two years); (5) malignancy of induced neoplasms; and (6) genotoxicity, measured in an appropriate battery of tests. Squire did not rank solvents in his published paper, but using his method of scoring, a ranking has been

TABLE 2
RISK RATIO (RR) FOR SOLVENTS (HAMMOND, 1979)

	ER/TLV	RR
PETROLEUM ALIPHATICS		
n-Hexane	1000/100*	10.00
Stoddard Solvent	10/100	0.10
ALCOHOLS		
Methyl Alcohol	610/200	3.10
Ethyl Alcohol	450/1000	0.34
Isopropyl Alcohol	100/400	1.25
AROMATICS		
Benzene	630/10	63.00
Toluene	240/100	2.40
Ethyl Benzene	91/100	0.91
CHLORINATED		
Carbon Tetrachloride	1280/5	256.00
Methylene Chloride	2750/200*	13.75
Trichloroethylene	620/100*	6.20
Perchloroethylene	280/100*	2.80
KETONES		
Methyl Butyl Ketone	87/25*	3.50
Methyl Isobutyl Ketone	165/50*	3.30
Methyl Ethyl Ketone	575/200	2.86
Acetone	1160/750	1.55
ACETATES		
Ethyl Acetate	615/400	1.54
n-Butyl Acetate	100/150	0.67
MISCELLANEOUS		
Carbon Disulfide	2260/10	226.00

*TLV Subsequently Lowered

TABLE 3

RANKING OF SOLVENTS SHOWN TO BE CARCINOGENIC IN AT LEAST ONE ANIMAL SPECIES

Solvent	Score*	Rank*
Chloroform	65	III
Dioxane	65	III
Carbon tetrachloride	40	IV
1,2-Dichloroethane	50	IV
2-Nitropropane	35	IV
Styrene	30	IV
1,1,2,2-Tetrachloroethane	40	IV
Benzene	25	V
Perchloroethylene	21	V
Hexachloroethane	25	V
Control - VCM	90	I

*Scores and Ranks as used by Squire, 1981

developed for several solvents that have been implicated as animal carcinogens (Table 3). The score and rank for vinyl chloride are included as a comparison. None of these solvents is highly carcinogenic by the Squire method. This system was developed to discriminate between the relative odds of risk, so that the government can make informed decisions on what is most hazardous to public health.

I have already indicated that few new health standards are expected under OSHA during the Reagan Administration because of OSHA's new policy on the requirements for a Standard and because of the Administration's commitment to regulatory reform, i.e., reduction of the burden of Federal regulations on business. But some health standards are bound to be promulgated. How can we tell ahead of time what new health standards might result? There are several indicators, and these are summarized in Table 4. Only the last two actions are direct indicators. The others are indirect, usually indicating new toxicity data or epidemiologic studies on substances that OSHA may eventually have to regulate. Solvent examples are available for most of the indicators.

In closing, I would like to say that the Occupational Safety and Health Act of 1970 is still a good act. It is a comprehensive law, covering much more than just standard setting and enforcement. We may not have frequent promulgation of new health standards; however, a number of health standards are already on the books and being enforced, there is a provision in the Act for emergency temporary standards, and the general duty clause can be utilized in certain situations. Moreover, I am not sure that we need standards for every conceivable hazard encountered in the workplace. There will no doubt continue to be challenges to our legislation and proposals for change. One recent legislative proposal is especially attractive--HR 638 introduced by Representative William A. Wampler (R-Va), which would establish a National Science Council for peer review of toxicologic findings made by regulatory agencies. Rep. Wampler had saccharin and nitrites in mind, but the same need exists for peer review of the science base of OSHA standards--generally considered one of the weakest parts in the standard setting process. Ending on a positive note, I should point out that the ACGIH TLV Committees can continue to be relied on in the U.S. and elsewhere to provide the best professional advice available for controlling chemical substances and physical agents in the workroom environment.

REFERENCES

1. Tabor, M. (1981). *Occupational Health and Safety* 50(12), 43-50.

2. Storch, W. J. (1981). *Chemical and Engineering News* 59(18), 36-39.

3. Hammond, J. W. (1982) Personal communication.

4. Scala, A. (1982). *Regulation* 6(Jan.-Feb.), 19-21.

5. Auchter, T. G., Assistant Secretary of Labor for Occupational Safety and Health (September 23, 1981). Statement before the Subcommittee on Investigations and General Oversight and the Subcommittee on Labor of the Senate Committee on Labor and Human Resources.

TABLE 4
EARLY INDICATORS
OCCUPATIONAL HEALTH STANDARD SETTING IN U.S.

ACTIONS	AUTHORITY	AGENCY/ORGANIZATION
(1) Requests for Information on Compounds for TLV List	Voluntary	ACGIH
(2) Proposed Rules	Clean Air Act, Amendments of 1977	EPA
(3) Proposed Test Rules	Toxic Substances Control Act: §4	EPA
(4) Notices of Substantial Risk	Toxic Substances Control Act: §8(e)	EPA
(5) Nomination of Chemicals for Testing	Secretary, DHEW (DHHS)	NTP
(6) Annual Reports on Carcinogens	Community Mental Health Centers Act, Amendment of 1978	NTP
(7) Health Hazard Evaluations	Occupational Safety & Health Act	NIOSH
(8) Current Information Bulletins	Occupational Safety & Health Act	NIOSH
(9) Candidate and Priority Lists of Potential Carcinogens*	Occupational Safety & Health Act	OSHA
(10) Advance Notices of Proposed Rulemaking	Occupational Safety & Health Act	OSHA

*Stay proposed by OSHA

6. Lemen, R. A., Director, Division of Standards Development and Technology Transfer, National Institute for Occupational Safety and Health (October 13, 1981). Statement on Priority Setting and Quantitative Risk Assessment to the Committee on the Institutional Means for Assessment of Risks to Public Health, the National Research Council Assembly of Life Sciences, Washington, D.C.

7. Hammond, J. W. (1979) Lecture notes.

8. Squire, R. A. (1981). *Science* 214, 877-880.

THE FEDERAL GERMAN APPROACH TO SETTING STANDARDS

J.K. Konietzko

*Institute for Industrial and Social Medicine
Johannes Gutenberg University,
Obere Zahlbacher Str. 67, 6500 Mainz/FRG*

HISTORY

Please forgive me if I sound a bit chauvinistic, but I believe that the German lists of limiting values are the oldest in the world. K.B. LEHMANN published the first maximum tolerance values for harmful substances at the workplace in 1886 (6). Through the years, this list was expanded until, in 1938, limiting values were available for approximately 100 harmful substances. The reasons for their inclusion were based on the results of toxicologic experiments and pharmacokinetic models as well as on empirical observations. Although these reasons were incomplete and, corresponding to the state of knowledge at that time, vinyl choride, for example, was classified as the least dangerous chlorinated hydrocarbon, they nevertheless represent an important milestone.

This work was interrupted by World War II and then resumed in 1955 when the German Society for the Advancement of Scientific Research (DFG) appointed a commission to test harmful substances at the work-place. Since there had been no opportunity for preliminary study, the commission adopted the US TLV lists. As of 1969, the maximum workplace concentrations or MAK values have been determined, validated, and published by the MAK Commission (3).

In 1972, a commission was appointed by the Federal Minister of Labor and Social Affairs to study harmful substances at the work-place. The commissions agreed that the limiting values for noncarcinogenic substances should be established by the MAK Commission and the recommended technical concentrations for carcinogenic substances, by the Ministerial Commission on Harmful Substances at the Work-Place. The reason for this unusual distribution of duties will be discussed in just a moment, when we take a closer look at the organization and mode of operation of both commissions.

MAK COMMISSION

The commission appointed by the DFG to test harmful substances at the work-place, the MAK Commission, is made up of 37 scientists, approximately half of whom come from the university setting and half from industry.

The commission meets annually to review the general situation and to determine future strategy. The preliminary research for this annual meeting is done in six study groups:
- Establishment of MAK values,
- Analytical chemistry,
- Occupational cancer,
- Determination of limiting values for dusts,
- Skin injury, and
- Establishment of limiting values in biological materials.

The determination of the MAK values and the classification of carcinogenic substances are based solely on toxicologic criteria; technical or economic aspects are not considered (3). The commission also attempts to validate all values scientifically; the results are made available to the public in a monograph series entitled "The Validation of MAK Values for Toxicology and Industrial Medicine" (2). This series has proved to be a valuable reference work for toxicologists.

The commission publishes two categories of limiting values: MAK values for the control of air at the work-place and BAT values for control of pollutants and their metabolites in biological material.

MAK VALUES

The MAK value is defined as the maximum permissible concentration of a harmful substance (gas, vapor, or material suspended in the air) at the work-place which, as far as we know, does not impair health nor annoy workers during repeated and long-term exposure (usually 8 hours a day, 40 hours a week). As a rule, the MAK value is the mean value for periods of up to one workday or one shift (9). The definition of the MAK value therefore closely approximates that of the US TLV-TWA. The MAK list, in contrast to the US TLV-STEL and TLV-Ceiling, does not, however, give peak concentrations for short-term loading (13); the inclusion of these values is currently under consideration (5).

THE FRG APPROACH TO STANDARD SETTING

The 1981 list included approximately 340 MAK values expressed in ppm or mg/m^3. Vapor pressure is also given for a few highly volatile substances. The letter "H" designates danger of absorption through the skin and "S", the danger of sensibilization (9). MAK values were not established for carcinogenic substances at the work-place, because the reversibility of the harmful effects is assumed in the scientific validation, at least at lower concentrations. Experimental evidence tends to indicate that this, however, is not the case with carcinogenic substances. Corresponding to the MAK definition, the commission does not cite limiting values. This does not imply that contact with such materials should be avoided, but rather that limiting values for carcinogenic substances cannot be scientifically validated.

Carcinogenic materials are classified into two groups: A 1, materials which can induce malignant tumors in man; and A 2, materials which have been shown to be potent carcinogens only in animal experiments under conditions comparable to the exposure of man at the work-place. Suspicious materials have been included in a third group (B); the possible carcinogenic potential of these substances has not yet been established. The previously valid MAK values therefore are continued until the carcinogenicity has been clarified. Suspected carcinogens recently accepted in the list are not assigned MAK values (9).

BAT VALUES

The harmful biological substance-tolerance value, the BAT value, is defined as the maximum permissible quantity of a harmful substance or its metabolites in man, or the subsequent deviation of a biological indicator from the norm, which, as far as we know, does not impair employee health, even with regular exposure at the work-place. Like the MAK values, the BAT values are based on a harmful substance load of 8 hours a day, 40 hours a week. BAT values, intended as maximum values for healthy individuals, are usually determined for blood or urine (7, 9).

The BAT values are justified, because many marginal conditions at the work-place such as different and discontinuous technical processes or interindividual differences (respiratory minute volume; differences in personal hygiene, skin absorption, biotransformation and elimination of foreign substances; interactions with drugs or alcohol) can modify regular dose-response relations produced under standardized laboratory conditions, which serve as the basis for the MAK values. In addition to external loading, the BAT value also considers the actual individual intake of foreign substances in the organism and their effect on the organism. It therefore is extremely well suited for estimating the health risk of the individual person. The fact that BAT values are currently available for only three substances, i.e., trichloroethylene, toluene, and lead, illustrates the methodological and conceptual difficulties involved in determining such limiting values. Nevertheless this concept represents an important step toward improved safety at the work-place.

TRK VALUES

There are a number of carcinogenic substances which, at least at present, are indispensable for industry. The MAK Commission does not help the safety inspectors by not providing limiting values. The Federal Ministry of Labor and Social Affairs appointed a commission on harmful substances at the work-place to provide guidelines and bases for safety in this situation. This ministerial commission establishes the recommended technical concentrations (TRK) for carcinogenic substances. It is made up of 34 experts from various fields of science, government agencies, unions, and industry. The double membership and flow of information guarantee good cooperation between the MAK Commission and the Ministerial Commission on Harmful Substances at tne work-Place.

The recommended technical concentration is that concentration of a harmful substance at the work-place (gas, vapor, or material suspended in the air) which serves as a guide for safety measures and analytical control at the work-place. These concentrations are oriented on the technical conditions and possibilities of technical prophylaxis utilizing experiences with harmful substances acquired in the field of industrial medicine. These are limiting values which cannot exceed the expectancy values of the concentration for one year, and limit the duration of exposure to no more than 8 hours a day, 40 hours a week (9, 12).

The relevant criteria for determining the TRK values are totally different from those for the MAK values. First of all, the pollutant concentration, measurable with analytic techniques, must be within the range of the TRK values, that is, it must be within the detection limits. The actual technical condition is decisive for the TRK level, in other words, how low can work-place concentrations be held, now and in the near future, using all available technical measures and methods of ventilation? This purely technical limiting value is then accepted or rejected, depending on

THE FRG APPROACH TO STANDARD SETTING

the results of experimental toxicologic tests and clinical experience in industrial medicine. 23 TRK values have now been established (12).

In short, one could say that the TRK values supplement the MAK values and that the TRK values pragmatically moderate the strict scientific criteria for the MAK values.

SCIENTIFIC RATIONALE OF LIMITING VALUES

This brings us to the most important question, "Which materials and criteria are used to validate limiting values?"

The MAK validations are based on current scientific publications which are accessible to the public (3). In addition, the MAK Commission also establishes and supports research projects.

The old dose-response relation, a concept which originated with PARACELSUS, forms the scientific basis for the MAK values. Such information, in principle, can be obtained in vivo and in vitro. The MAK Commission, however, basically prefers experiences in man to those obtained in animal tests. The real situation is different. A few casuistic reports on accidents with high concentrations are available for several substances. Few tests with volunteers are available which were carried out under standardized conditions at low concentrations. The few epidemiologic investigations on larger exposed collectives are full of methodological errors. These investigations must constantly fight the methodological errors arising in the work-place due to frequent personnel changes, different methods of operation, and different harmful substances. It therefore is extremely difficult to analyze an adequately large, basically homogenous collective with well verifiable and comparable exposure conditions. In most cases, the epidemiologic results provide only a qualitative "yes" or "no". Quantification of the dose-response relation is rare. Tests using volunteers, even if well planned and safe, still raise ethical questions and often cannot be carried out for psychological reasons.

Even in the future, in vivo studies will be supported by animal tests, the main problems of which are well known and have often been discussed (14). While individual results of animal experiments are often applicable to man, a definitive answer to this question is impossible. Animal tests obviously cannot be used to analyze subjective factors such as feeling of ill health, nasal nuisances, and similar annoyances considered in the MAK list. These tests are far more suitable for evaluating behavioral disturbances induced by the acute effect of volatile, fat-soluble substances on the central nervous system and have been applied in this field with good results. They are best used to examine organically detectable processes, i.e., pharmacokinetics and biochemical or histologic demonstration of toxic organ effects. In principle, animal tests, even in this area, can only answer those questions which the experimenters set for the test (14). They, after all, are the ones who must consider the species-specific physiologic requirements in man and in animals and must be aware that species-specific distribution and metabolic processes can lead to qualitatively and quantitatively different results. False interpretations will continue to be made until more is known about the physiologic requirements in experimental animals, particularly the enzymatic potential. This fact should be remembered when attempting to estimate the applicability of animal experiments to man. With increased knowledge of these requirements and improved selection of experimental animals, animal tests will provide a better basis for the estimation of health risks in man. In the future, multiple loads at the work-place and interactions with pollutants from the private sphere, such as alcohol and drugs, will be evaluated with a reasonable degree of reliability.

Evaluating carcinogenic substances at the work-place involves a special set of problems. Most potent carcinogens have already been identified and eliminated from production processes or have been neutralized technically. Substances still suspected of being carcinogenic are usually weak carcinogens. To establish the carcinogenicity of these substances in animal experiments, long-term tests with many animals are necessary for mathematical and statistical calculations. Such investigations are both time-consuming and ex-expensive, taking up to 3 years and costing as much as $200,000 (1, 8). Consequently, we have to select short-term tests and then verify the results in animal tests. Several short-term tests are available. Mutagenicity tests on different strains of Salmonella typhimurium have been studied thoroughly and are now accepted. The common principle of such tests is that most organic carcinogens in the cells first attack the DNA, i.e., the genetic material. These substances can be both carcinogenic and mutagenic. Short-term tests then demonstrate DNA damage at the molecular, cellular, and multicellular level. We cannot deal with the methodology in any more detail here; let it suffice to say that a comparison of studies on carcinogenic and mutagenic effects of established organic carcinogens show a 88% to 94% correspondence (11).

THE FRG APPROACH TO STANDARD SETTING

IMPORTANCE OF LIMITING VALUES

Both MAK and TRK values are published annually in the "Bundesarbeitsblatt" by the Federal Minister of Labor and Social Affairs. These values, however, are only suggestions made by a group of experts; they have no legally binding character. This status also reflects the intention of the commission and prevents flexible adaptation to medical needs and new scientific knowledge from being sacrificed to the long process of legislation (3). My experience indicates that these values are not only a nonbinding moral instance, but they are also a very important standard for medical and technical safety measures at the work-place and a valuable help for settling damage claims and lawsuits in the Social Court. Observance of the limiting values can also be declared legally binding in situations such as licensing procedures dictated by laws regulating environmental nuisances and ordinances controlling the use of harmful substances at the work-place.

OPEN QUESTIONS

I would now like to mention two points which are still unclarified.
1. Lacking adequate information, the MAK Commission was not in a position to cite limiting values for pollutant mixtures. It was unable to adopt the US suggestion of simply adding the partial components together, because chemical and toxicologic requirements were absent (4). Instead, it attempts to clarify the questions with suitable toxicologic models. Considering the virtually unlimited combinations and the diversity of the products already commercially available, this problem can only be solved to a limited degree.
2. Almost no information is available on the tolerance of harmful substances at the work-place during pregnancy. Pregnant women therefore should not be employed at workplaces where such substances are being used. This, however, only solves part of the problem, since the fetotoxicity of chemical substances at the work-place can be most potent during the first two gestational weeks, i.e., before the woman knows she is pregnant. The suggestion that women of child-bearing age not be exposed to harmful substances at all or only to very low concentrations therefore is not altogether inappropriate.

SUMMARY

The MAK values, established and regularly revised by an independent commission of scientists, are based only on toxicologic criteria. They have no legally binding character, but they are usually employed in the plant setting as though they did. MAK values have not been established for carcinogenic substances on general principles. Should these substances prove irreplaceable for technical reasons, a commission is appointed by the Federal Ministry of Labor and Social Affairs to determine the recommended concentration, based on technical and medical criteria, that virtually excludes the possibility of a carcinogenic risk.

REFERENCES

1. Fox, J.L. (1977). *Chem. Eng. News*, Dec. 72, 34-46.
2. Gesundheitsschädliche Arbeitsstoffe (1981). "Toxikologische-arbeitsmedizinische Begründung von MAK-Werten" (Ed D. Henschler), Verlag Chemie, Weinheim.
3. Henschler, D. (1981). *In* "Wissenschaftliche Grundlagen zum Schutz vor Gesundheitsschäden durch Chemikalien am Arbeitsplatz", pp. 29-40. Harald Boldt Verlag, Boppard.
4. Henschler, D. (1981). *In* "Wissenschaftliche Grundlagen zum Schutz vor Gesundheitsschäden durch Chemikalien am Arbeitsplatz", pp. 92-96. Harald Boldt Verlag, Boppard.
5. Henschler, D., zur Mühlen, Th., Drope, E. (1979). *Arbeitsmed.,Sozialmed., Praeventivmed*,14, 191.
6. Lehmann, K.B., Flury, F. (1938). "Toxikologie und Hygiene von technischen Lösemitteln". Julius Springer Verlag, Berlin.
7. Lehnert, G. (1980). *Arbeitsmed., Sozialmed., Praeventivmed.*, 11, 266-270.
8. Maugh, T.H. (1974).*Science* 183, 94.
9. "Maximale Arbeitsplatzkonzentrationen 1981. Harald Boldt Verlag, Boppard.
10. Norpoth, K. (1981).*In* "Wissenschaftliche Grundlagen zum Schutz vor Gesundheitsschäden durch Chemikalien am Arbeitsplatz", pp. 41-48. Harald Boldt Verlag, Boppard.
11. Purchase, I.F.M., Longstaff, E., Ashby, J., Styles, J.A., Anderson, D., Lefevre, P.A., Westwood, F.R. (1976). *Nature* 264, 624-627.
12. Schütz, A. (1981). *Moderne Unfallverhütung* 24, 30-33.
13. TLVs, Threshold Limit Values for Chemical Substances in Workroom Air, Adopted by ACGIH for 1981. Publications Office ACGIH, Cincinnati, OH.

THE UNITED KINGDOM APPROACH
TO THE SETTING OF STANDARDS

C D Burgess

Health and Safety Executive
25 Chapel Street, London NW 1 2DT

INTRODUCTION

If a standard for exposure to a toxic substance is to be more than a mere exhortation to limit exposure to a given value, the standard needs to be given suitable status. It cannot stand alone. This status is usually provided by linking the standard, to a greater or lesser extent, to legislation.

Standards for the control of exposure to toxic substances have been incorporated directly in UK legislation for many years. For example, in the *Indiarubber Regulations 1922 (1)*, no "fume process" (which is a process using carbon disulphide, sulphur dichloride, benzene, carbon tetrachloride or trichlorethylene) may be carried on without the use of an efficient exhaust draught ... which prevents the dust or vapour from entering air of any room in which persons work". This is a very stringent standard indeed. No doubt it served its purpose at the time but construed strickly it is neither achievable nor enforceable. It would not be enacted in such a form today.

Linking standards to legislation inevitably gives rise to problems. As a target to be achieved, a standard can be wholly based on the medical effects. On the other hand, when a standard is linked to legislation, much more detailed consideration needs to be given to its interpretation; in particular to methods of sampling and analysis, the period over which the average concentration should be measured its relation to other standards such as biological limit values, criteria for compliance and non-compliance, action to be taken in the event of the standard being exceeded, and the resources available both in human and financial terms, in establishments both large and small, to meet the requirements of the standard.

2. RELATIONSHIP BETWEEN STANDARDS AND LEGISLATION

In the United Kingdom there are three basic options, all of which have been used in the past but as there are few examples which involve solvents these options will be illustrated by reference to a wider range of toxic substances.

(a) Standards incorporated directly in legislation

If a standard is incorporated directly in legislation there is a choice between making compliance with it an absolute duty or qualifying the requirements by "so far as is reasonably practicable" which allows costs to be weighed against risks. *The Ionising Radiations (Sealed Sources) Regulations (2)* is an example of the former. Regulation 11 requires "... no person shall receive any radiation dose in excess of those permitted under the ... regulations". However, in the case of toxic substances the incorporation of standards directly in legislation is an option which is not often used.

(b) Standards incorporated into an Approved Code of Practice or an Approved List in support of a general legislative requirement

The concept of an Approved Code of Practice giving practical guidance on legislative requirements was introduced by the *Health and Safety at Work etc Act 1974 (4)*. If a person fails to comply with a standard incorporated in an Approved Code of Practice the burden of proof is then transferred to him to show that he has complied with the legislative requirements in some other way. *The Approved Code of Practice (5)* which gives practical guidance on the *Control of Lead at Work Regulations (6)* is a recent example of the incorporation of a standard in an Approved Code of Practice. The regulations place a duty on the employer to provide "adequate control" of the exposure of his employees to lead. "Adequate control" is defined in the Approved Code of Practice by reference to the lead in air standard.
Another illustration of this option is the use of a list approved by the Health and Safety Commission (HSC). This concept has been used in the *Dangerous Substances (Conveyance by Road in Road Tankers and Tank Containers) Regulations (7)*. The approach is also intended to be used for the proposed *Classification, Packaging and Labelling Regulations,*

THE UK APPROACH TO STANDARD SETTING

the *Consultative Document (8)* for which has recently been issued. These lists contain data such as warning symbols, risk phrases and safety phrases, not exposure limits, but they do illustrate a useful method of providing a status for standards which is closely related to legislation and yet can be amended without the long delays associated with making regulations. This method is likely to be increasingly employed in future.

(c) General legislative requirements supported by standards in guidance literature

A further option is to use an exposure standard as an authoritative but non-statutory interpretation of general legislative requirements by publishing the standard in guidance literature. Apart from the advantages of this procedure in its own right, it is very useful as a temporary measure pending more long term action. This approach has been used by the Health and Safety Executive (HSE) in relation to the *Asbestos Regulations (9)*, and for *acrylonitrile (10)*. It has also been used rather more generally in relation to a wide range of substances (*11* and *12*).

In most cases the underlying statutory requirement is Section 2 of the HSW Act which requires every employer to ensure, so far as is reasonably practicable, the health, safety and welfare at work of all his employees.

3. SELECTING VALUES FOR STANDARDS

Given the existing legislative framework in Britain, particularly the general requirements of the HSW Act, there is a need to provide clear and authoritative guidance by way of standards for employers, employees and their safety representatives, and enforcing authorities on what levels of exposure may be deemed to satisfy an employer's obligation under the relevant statutory provisions.

(a) Type of limits

Most exposure standards set in Britain relate to airborne concentrations of toxic substances. Since most individual intakes occur in this way, this is clearly a useful parameter. Such standards are objective but have both the advantage and disadvantage of being independent of an individuals metabolism. Exposure limits based on airborne concentrations provide an unequivocal basis for the design of equipment and for checking compliance but they do not provide a measure of the ill effects to a particular individual.

But many solvents are readily absorbed through the skin and where this route of entry is significant, clearly reliance cannot be placed wholly on airborne limit values. The problem is exacerbated when, as is often the case, absorption by inhalation and through the skin occurs simultaneously. Due to the practical difficulties of developing quantitative standards for the control of skin absorption, reliance is usually placed on the provision of an effective barrier by way of protective clothing between the individual and the contaminant.

Research into biological monitoring is carried out at the Health and Safety Executive's Occupational Health and Medicine Laboratories at Cricklewood. Guidelines for action levels for several metabolities of toxic substances are published in Guidance Notes. Nevertheless, with the exception of lead, biological limit values have not so far been directly linked to legislation and the emphasis is still directed to limit values for airborne contaminants.

(b) Methodology

Standards which are considered to be reasonably practicable for particular substances are recommended to the Health and Safety Commission (HSC) by its Advisory Committee on Toxic Substances (ACTS) which consists of representatives of employers, employees, local authorities, scientific and medical experts. This objective entails assessing on the one hand, the nature of the hazard and the extent of the risk and on the other, the consequences involved in reducing the risk.

Toxicity reviews *(13)*, data on current exposure levels, the consequences involved in making changes and methods of sampling and analysis are prepared for the Committee by HSE. An insight into the logic employed is given in the *Final Report of the Advisory Committee on Asbestos (18)*. Essentially the approach is to identify the lowest level of exposure which can be achieved throughout industry and compare that level with the available dose response data. If the risk is unacceptable at that level and exposure cannot be reduced then consideration has to be given to banning those processes taking into account the social consequences of such action. By an interative process a level of exposure is obtained where the level of risk is acceptable and where further expenditure of effort in the processes where control is most difficult is out of all proportion to the to the reduction of risk thereby achieved.

If however some residual risk is still present, as it may well be in the case, for example, of carcinogens, the legal obligation to reduce exposure levels still further in those processes where it is reasonably practicable to do so, still applies. These standards therefore represent the upper limit of permitted exposure, not a safe level such that once attained, no further improvement in control is necessary.

(c) Technical and economic feasibility

Standards recommended by ACTS to the HSC are what the Committee consider to be "reasonably practicable" for that substance. While the selection of values for standards is thus based largely on the medical and scientific evidence, care is also taken to see that there is not a gross disproportion between the risk and the costs involved in the measures necessary for averting it. This factor appears to distinguish standards agreed by the Health and Safety Commission from those published by *Deutsche Forschungsgemeinschaft (15)* which specifically state that "scientifically based criteria for health protection, rather than their technical or economically feasibility, are employed" and from those in the *ACGIH list (16)* which is silent on this issue.

Exposure standards based solely on medical effects might hopefully be expected to obtain a reasonable measure of international agreement (although even in this case, variations may be expected due to local factors). International agreement on standards which take into account technical and economic feasibility are likely to be more difficult to achieve. Even in the European Communities (EC) where one would expect these factors to be fairly similar, difficulty is still being experienced in reaching agreement on standards for lead *(17)* which, since they will be mandatory, must take technical and economic considerations into account.

4. SOME TECHNICAL ISSUES

Several technical issues arise in the standard setting process. Their relative importance varies according to the beholder's functions in relation to the application of standards. For airborne limit values probably two of the most important are methods of sampling and analysis and time averaging.

(a) Methods of sampling and analysis

Clearly those concerned with standard setting need to make it quite clear whether the values selected are intended to relate to an individual intake, the value obtained from a personal sampler, the workroom concentration or to some other parameter.

In the United Kingdom in the past practice has tended to vary. For example, in setting the hygiene standard for cotton dust the Joint Standing Committee in 1973 recommended that "an acceptable airborne dust concentration for workrooms ... shall be up to 0.5 mg/m^3 of total cotton dust less fly".

On the other hand, the *Control of Lead at Work Regulations (6)* requires "adequate monitoring procedures to measure the concentrations of lead in air to which employees are exposed". The *Code of Practice (5)* clearly intends that personal sampling should be the norm although static sampling can be used subject to certain conditions. The latest draft of the *European Communities Directive on Lead at Work (17)* contains in Article 4 the phrase "all lead in air measurement shall be representative of worker exposure to airborne particles and/or aerosols containing lead". A praiseworthy objective but easier to state than to achieve. The *Council Directive on the protection of workers exposed to vinyl chloride monomer (19)* defines the "technical long term limit value as "vinyl chloride monomer in the atmosphere of a working area". It requires that all working areas should be monitored and the points chosen should be as representative as possible of the exposure level of workers in that area. Continuous monitoring is obligatory for enclosed VCM polymerisation plant.

When the HSC approve standards, it is the HSE's aim to publish appropriate methods of sampling and analysis *(14)* to complement the standards. It is envisaged that the development of improved methods for personal sampling such as the thermal desorption passive sampling tube which is now commercially available for organic vapours *(20)* will result in increasing emphasis being placed on personal sampling.

(b) Time averaging

The choice of the optimum averaging time for the exposure limit set out in the standard is a difficult one. Ideally, the averaging time ought to be related to the toxic effects of the substance. Chronic effects might merit

THE UK APPROACH TO STANDARD SETTING 187

averaging times perhaps in the order of years. For acute effects, limits related to a few minutes might be more appropriate. But the practical problems of measurement need to be taken into account. No enforcing authority could happily contemplate obtaining evidence for enforcement action on a standard averaged over 12 months. An employer too, is placed in grave difficulties if he has to comply with a "ceiling value", the averaging time for which is strictly infinitesimally short. The concept of short term exposure limits as defined in the *ACGIH list (16)* also presents practical difficulties.

Averaging times in standards linked to legislation need to be simple in concept and capable of practical application throughout industry. For these reasons ACTS is currently considering proposals which contain only 2 variants of averaging times. For most substances giving rise to chronic effects a standard 8 hour averaging time would be used, although whether 8 hours is now an appropriate period for full shift sampling times is open to question. For substances with acute effects, the standard would relate to a shorter averaging time of the order of 10 minutes or less depending on the toxic effects of the substance.

(c) Compliance and sampling strategies

Some work has been done in HSE *(21)* on the statistical principles which underly criteria for compliance. Most of it has been related to the VCM limit value set out in the EC Directive *(19)* because this was agreed in the form of an annual limit value which is difficult to enforce rather than an 8-hour TWA. This is a fairly complex problem because employers are concerned with compliance and enforcing authorities with non-compliance. Detailed statistical guidance on compliance has been deliberately felt flexible at this stage. Guidance on sampling strategies has however been issued for particular substances such as lead *(22)* and reference has already been made to publications on methods of measurement *(14)*.

(d) Action if standards are exceeded

Exposure significantly in excess of an exposure limit would normally result in a contravention of the relevant legal requirement. While the basic requirement to reduce exposure to the lowest level that is reasonably practicable would continue to apply, only in a few cases are there explicit requirements which state specifically what action should be taken. For example, in coal mines (3) if the respirable dust index exceeds the permitted amount, the inspector must be notified about what action is being taken unless work is stopped. If the dust index continues to exceed the permitted amount of work must be stopped. In the case of the Control of Lead Regulations there is the back up provided by a biological standard and the requirement to suspend an employee from work if this biological standard is exceeded. The EC Directive on lead currently under negotiation is on the same lines although expressed rather differently there being a requirement that work is not to continue in an area unless measures, for example, the wearing of respiratory protective equipment, are taken.

The Asbestos Regulations only create an offence and are silent on what further action should be taken if the standard is exceeded. The EC Directive on asbestos currently being negotiated would require all necessary steps to be taken to remedy the situation as soon as reasonably practicable and the issue to worker of personal protective equipment which they would have a duty to wear. The draft directive also calls for sampling and analysis to be repeated within a period of 8 days. Where the limit values are still being exceeded, it would require all work in the area to cease until the situation is remedied.

5. FUTURE TRENDS

The *Health and Safety Commission's plan of work for 1981-82 and onwards (23)* sets out proposals for the control of substances hazardous to health.

The rationalisation of existing controls in a legislative framework which is more or less universally applicable to the control of all toxic substances is being considered. This is particularly appropriate since the EC adopted the *Directive on the protection of workers from the risks of exposure to chemical, physical and biological agents (25)* and from which it may be anticipated that daughter directives dealing with specific substances may stem. Preliminary proposals for regulations for the control of substances hazardous to health have been prepared and will be considered by the ACTS. These regulatory proposals set out the principles of occupational hygiene in legislation form and should, because they would apply a uniform regime to all substances hazardous to health, considerably simplify the task of employers in complying with their statutory obligations, of safety representatives in fulfilling their statutory functions, and of competent authorities in their enforcement duties. These proposals are also intended to avoid the time consuming problem of producing new regulations to implement each EC directive as it is adopted.

It is intended that exposure standards should be linked to these regulations. As a first step proposals for a revision of guidance note EH "threshold limit values" have been put to the ACTS. It is proposed that the HSE should publish its own list of exposure limits in three parts. Part 1 would contain control limits which are either included in regulations or are closely associated with them and those which have been adopted by the HSC on the recommendation of the ACTS. Part 2 would contain a list of limits which the HSE considers should not be exceeded but which have not been formally adopted by the HSC. These limits would provide clear guidance for industry and would be used by HSE inspectors in assessing compliance with the relevant statutory provisions. Part 3 would contain a list of those substances for which it is proposed, within a given time, to change the existing limit or to introduce a limit for the first time. This would provide adequate warning to all those likely to be affected and given time for comments to be made and, where necessary, adaptation to be made to processes.

6. CONCLUSIONS

In the past 20 years or so there have been significant developments both in the philosophy of the control of exposure to toxic substances and in the equipment available for assessing exposure. This has resulted in marked improvements in industrial conditions in which standards for exposure have played an important part. It seems likely that new equipment may significantly increase the precision and reduce the expense of monitoring but there is a risk that the philosophy may become too complex for the average employer and employee to understand. It is an appropriate moment for review and rationalisation to ensure that legislation and guidance on the control of toxic substances is comprehensible to all, achievable by employers, acceptable to employees and enforceable by the authorities.

REFERENCES

1. *The Indiarubber Regulations, 1922.* S.R.&O. 1922, No 329 as amended by S.I. 1973. No.36. HMSO. London.

2. *The Ionising Radiations (Sealed Sources) Regulations, 1969.* S.I. 1969, No.808, as amended by S.I. 1973, No.36. HMSO. London.

3. *The Coal Mines (Respirable Dust) Regulations, 1975,* S.I. 1975. No. 1433 as amended by S.I. 1978. No. 807. HMSO. London.

4. *Health and Safety at Work Etc Act 1974 s.s. 16-17.* HMSO. London.

5. Health and Safety Commission. (1980). *Control of Lead at Work. Approved Code of Practice.* HMSO. London.

6. *The Control of Lead at Work Regulations 1980.* S.I. 1980 No. 1248. HMSO. London.

7. *The Dangerous Substances (Conveyance by Road in Road Tankers and Tank Containers) Regulations. 1981.* S.I. No. 1059. HMSO. London.

8. Health and Safety Commission. (1981). *Proposals for Regulations for the Classification, Packaging and Labelling of Dangerous Substances. Consultative Document.* HMSO. London.

9. Health and Safety Executive. (1976). *Asbestos. Guidance Note EH 10.* HMSO. London.

10. Health and Safety Executive. (1981). *Acrylonitrile. Guidance Note EH 27.* HMSO. London.

11. Health and Safety Executive. (1977). *Toxic substances: a Precautionary Policy Guidance Note EH 18.* HMSO. London.

12. Health and Safety Executive. *Threshold Limit Values 1980 Guidance Note EH 15.* HMSO. London.

13. Health and Safety Executive. *Toxicity Review - Carbon Disulphide.* HMSO. London.

14. Health and Safety Executive (1981). *Methods for the Determination of Hazardous Substances.* HMSO. London.

15. Commission for Investigation of Health Hazards of Chemical Compounds in the Work Area (1981). *Maximum Concentrations at the Workplace.* Report No XVII Deutsche Forschungsgemeinschaft. Bonn.

16. American Conference of Governmental Industrial Hygienists. (1981). *Threshold Limit Values.*

17. *Proposals for a Council Directive on the Protection of Workers from risks related to exposure to lead at work* O.J. No. C324, 28.12.79. p.3. Commission for the European Communities. Brussels.

18. Health and Safety Commission (1979) *Asbestos Final Report of the Advisory Committee.* Vol. 1, p.p. 72-73. HMSO. London.

19. *Council Directive on Protection of Workers from Vinyl Chloride Monomer.* O.J. No. L197, 22.7.78. p.12. Commission for the European Communities. Brussels.

20. Brown, R.H., Walkin K.T. (1981). *Proc. Roy. Soc. Chem. (An. Div.)* 18, 205-208.

21. Harvey R.P. (1981). *The Application of Statistical Principles in the Measurement and Assessment of Exposures to Airborne Contaminants.* Conference on the Evaluation of Airborne Contaminants. Oyez Communications Group. London.

22. Health and Safety Executive (1981). *Control of Lead: air sampling techniques and strategies. Guidance Note EH 28.* HMSO. London.

23. Health and Safety Commission (1981). *Plan of Work 1981-82 and onwards.* HMSO. London.

24. *Council Directive on the protection of workers from the risks of exposure to chemical, physical and biological agents.* O.J. No.L.327. 3.12.80. p.8. Commission for the European Communities. Brussels.

USE OF DETECTOR TUBES FOR EVALUATING MIXED SOLVENT VAPOURS

J.A. WALTON
Draeger Safety, Sunnyside Road
Chesham, Bucks

INTRODUCTION

Much of what has been written in the past about the detection and measurement of toxic gases and vapours and many of the devices designed for such measurements assume that only a single pollutant is present. In practice, many solvents used in industry are not single compounds but mixtures. Sometimes, for example paint solvents, these mixtures are deliberate formulations designed to give the desired properties. On other occasions, the solvent is a distillation fraction that can contain a range of compounds.

If one or more of the components in such a mixture is known or thought to be toxic it is necessary to provide some form of monitoring. The choice of area monitoring, personal monitoring, continuous monitoring or grab sampling is described by Thain (1) and it is not the subject of this paper. This paper describes only the use of Detector Tubes.

When trying to monitor mixtures, we can be faced with one of several problems:-

1. to determine every component from the mixture

2. to determine a single component from the mixture

3. to determine the total amount of contamination from the mixture without separating the components

Naturally, any of these problems can be solved given enough laboratory time and the services of a skilled analyst. The purpose of this paper is to try and indicate methods, whereby quick, simple and approximate answers may be obtained.

Detector Tubes

Detector tubes are simple glass phials or ampoules filled with chemical reagents adsorbed onto a porous insert support material. The reagents are chosen to give a distinct colour change in the presence of a specific pollutant. Usually the length of the discolouration is a measure of the concentration of the pollutant but sometimes it is the intensity of the colour stain which has to be evaluated. The air is drawn through the tube with a simple hand-operated pump, usually a bellows pump. The whole principle is fully described by Leichnitz (2) and Walton (3).

Threshold Limit Values When monitoring the vapours of solvent mixtures, it is as well to bear in mind the criteria which will indicate whether the situation can be considered acceptable or not. Whilst threshold limit values, TLV's, have been established in the U.K. and U.S.A. and similar limit values in other countries:- M.A.K., T.R.K., M.A.C., etc., for single substances; no clearly defined TLV's exist for mixtures. Some guidance is offered in the list of TLV's published by the Health and Safety Executive in the U.K.(4). The choice of an appropriate limit value from a mixture will depend upon the nature of the mixture. It is not the intention of this paper to discuss these limit values for mixtures.

DELIBERATE MIXTURES

Example 1

The solution to the type of problem described in (1) above would depend very much upon the exact nature of the mixture. If it consisted of a few compounds of distinctly distinguishable chemical properties, e.g. Ethanol, Toluene, Trichloroethane, then the problem could be easily solved. Ethanol could be measured by means of Draeger Tubes Alcohol 100/a; Toluene with the tube Toluene 5/a; and Trichloroethane by means of tube Trichloroethane 50/d. That this particular problem is capable of such easy solution depends upon the fact that there is little cross-sensitivity on any of these three tubes by the other two compounds.

Example 2 As another example, we might consider the solvent mixture:- 60% Ethyl Acetate, 20% 2-propanol, 10% n-hexane, 10% Toluene.

The tube, Alcohol 100/a, will be specific for 2-propanol

and tube Toluene 5/a specific for Toluene. However, the tube Ethyl Acetate 200/a would indicate the propanol and n-hexane as well as the Ethyl Acetate, all with about the same sensitivity. This might suggest that this tube could be used to measure the total of these three solvents, unfortunately this is not the case, tube Ethyl Acetate 200/a indicates toluene with about ten times the sensitivity, i.e. 200 ppm toluene would read 2000 ppm on this tube. Similarly, the tube n-hexane 100/a would indicate all four substances with varying degrees of sensitivity.

Monitoring Techniques. There are three different ways of tackling this sort of problem. One method is to assume that to first approximation the proportions of the four vapours in the air will be the same as the proportions of the four compounds in the solvent. This is reasonably true if the compounds all have similar rates of evaporation. We know we can determine the propanol and toluene and the ratio of these two should be approximately 2:1. If this is the case, then we can estimate the ethyl acetate concentration as being three times the propanol concentration and the n-hexane concentration as being equal to the toluene concentration.

The second method is to undertake a series of tests using all four detector tubes alongside some laboratory method that will accurately measure all four components. In this way, a correlation will be obtained between the tube readings and the vapour concentration. Once this correlation is achieved then the detector tube may be used for quick routine checks. This idea will be expanded upon later

The third method is to use the Ethyl Acetate tube, which is known to indicate all four components, for a simple go/no-go test. If the acceptable criteria is laid down that the stain must not exceed "200" on the tube, then this would mean that the sum of the concentrations of the ethyl acetate, propanol, n-hexane plus ten times the amount of toluene would not exceed 200 ppm. This should be well inside the TLV for such a mixture.

DISTILLATION FRACTIONS

Other types of solvents are distillation fractions, as such these have no clearly defined chemical composition. This makes the calibration of a detector tube very difficult. One sample of solvent from a particular batch, from one supplier, will give a totally different result to that obtained from different sample. Some idea of the range of variation can be seen from the results obtained for "white spirits".

TABLE 1

Concentration of White Spirits ppm	Reading on Petroleum Hydrocarbons 100/a Tube ppm
100	75 - 225
200	150 - 450
400	200 - 600
800	350 - 1050
1600	500 - 1500

CALIBRATION OF DETECTOR TUBES FOR HIGH BOILING SOLVENTS

Many solvents used in industry are chosen, on the grounds of safety, for their high flash points. Typical solvents are:- SBP6, SBP11, White Spirit, EXSOL5, Halpasols.

These solvents consist in the main of long chain aliphatic hydrocarbons. In an attempt to approximate a calibration for these substances, we have undertaken calibrations for undecane ($C_{11}H_{24}$) and dodecane ($C_{12}H_{26}$) on two different detector tubes: Hydrocarbons 2, Petroleum Hydrocarbons 100/a.

Hydrocarbon 2 tube   A colour comparison tube.

Method: As the stain is darker than the given colour comparison layer, it is necessary to continue pumping until the entire indicating layer is discoloured.

Undecane:
Concentration: Approximately 530 ppm (3.5mg/litre) (corresponds to about 100% vapour pressure at 20°C, 1013 mbar)

Number of pump strokes required for complete discolouration of the indicating layer: 40

Dodecane:
Concentration: Approximately 200 ppm (1.5mg/litre) (corresponds to about 100% vapour pressure at 20°C, 1013 mbar)

Number of pump strokes required for complete discolouration of the indicating layer: 100

These results may be compared with the previously published data for Kerosene, Leichnitz (5). The tube - Hydrocarbons 2 is not really suitable for low concentrations of high boiling hydrocarbons.

Petroleum Hydrocarbon 100/a tube. A stain length tube
Method: Using the same saturated vapour atmosphere as described above, the tube reading was determined for a number of different pump strokes.

Undecane:
Concentration: 530 ppm

TABLE 2a

Number of Pump Strokes	Tube Reading
4	100
6	150
9	200

Dodecane:
Concentration: 200 ppm

TABLE 2b

Number of Pump Strokes	Tube Reading
10	100

The Petroleum Hydrocarbons 100/a tube at least offers some prospect of measuring the concentration of these high boiling solvents in the 100-500 ppm range.

QUALITATIVE INDICATORS

Sometimes it is not possible to obtain a quantitative result using a detector tube but it may still be possible to obtain a qualitative indication of the presence of the vapour in the atmosphere.

Certain detector tubes use reagent systems of such a general nature that they will react with a wide range of substances. Two such tubes are the Polytest tube and the Ethyl Acetate 200/a tube.

Polytest Tube

The reagents in the Polytest tube are iodine pentoxide and fuming sulphuric acid. A large number of reducing substances will react with these reagents to reduce the iodine pentoxide to brown iodine. However, depending upon the exact nature of the reacting substances and the reaction conditions it is possible to get reddish, brownish green or completely green reaction products (due to the formation of "iodine sulphate"). Chlorinated solvents can give rise to yellow stains - iodine chloride. Aromatic Hydro-carbons such as benzene will react with the fuming sulphuric acid to form colourless sulphonic acids. Fortunately, however, a part of the benzene present will react with the iodine pentoxide to give a brown stain.

When using a tube such as the Polytest tube for checking atmospheric contamination, it must always be remembered that lack of a colour stain does not necessarily indicate that the atmosphere is free from contamination. There are basically six different situations which can arise:-

1. No contamination — no stain

2. Contamination present which reacts in the tube to produce a colour — stain

3. Contamination present but which does not react in the tube — no stain

4. Contamination present which reacts in the tube without causing a colour change. — no stain

5. Contamination present in such high concentration as to "poison" the reagent in the tube or to "bleach"

DETECTOR TUBES FOR MIXED VAPOURS

any stain produced	— no stain or faint stain
6. More than one contaminant present such that one interferes with the indication of the other	— no stain or faint stain

Examples of substances which do react on the Polytest tube to give a colour stain are given in table 3.

TABLE 3

Substance	Concentration	Number of strokes of the bellows pump	Length of discolouration	Notes on the indication
Acetone	5000 ppm	5	approx. 10 mm	brownish green
Acetone	above liquid	5	completely coloured	brownish
Acetylene	200 ppm	5	approx. 10 mm	brownish green
Acetylene	high conc. (over 1%)	5	completely coloured	brownish
Arsine	10 ppm	5	approx. 10 mm	brownish green
Arsine	high conc. (over 1%)	5	completely coloured	brownish
Benzine (Gasoline)	50 ppm	5	approx. 10 mm	brownish green
Benzine (Gasoline)	above liquid	5	completely coloured	brownish
Benzene	100 ppm	5	approx. 10 mm	brownish
Benzene	above liquid	5	approx. 10 mm	brownish
Butane	100 ppm	5	approx. 10 mm	faded green (spotty)

TABLE 3 (contd.)

Substance	Concentration	Number of strokes of the bellows pump	Length of discolouration	Notes on the indication
Butane	high conc. (over 1%)	5	completely coloured	brownish green
Carbon disulphide	10 ppm	5	approx. 10 mm	greenish
Carbon disulphide	above liquid	5	completely coloured	brownish green
Carbon monoxide	100 ppm	5	approx. 10 mm	brownish green
Carbon monoxide	high conc. (over 1%)	5	completely coloured	brownish
Ethylene (ethene)	500 ppm	5	approx. 10 mm	brownish green
Ethylene (ethene)	high conc. (over 1%)	5	completely coloured	brownish
Nitrogen monoxide (NO)	50 ppm	5	approx. 10 mm	brownish green
Nitrogen monoxide (NO)	high conc. (over 1%)	5	completely coloured	brownish with a bleaching effect
Perchloroethylene	50 ppm	5	approx. 10 mm	greenish
Perchloroethylene	above liquid	5	completely coloured	brownish green
Propane	500 ppm	5	approx. 10 mm	faded green (spotty)
Propane	high conc. (over 1%)	5	completely coloured	brownish green
Styrene (monostyrene)	500 ppm	5	approx. 10 mm	brownish
Styrene (monostyrene)	above liquid	5	approx. 10 mm	brownish

DETECTOR TUBES FOR MIXED VAPOURS

TABLE 3 (contd.)

Substance	Concentration	Number of strokes of the bellows pump	Length of discolouration	Notes on the indication
Toluene	200 ppm	5	approx. 10 mm	brownish
Toluene	above liquid	5	approx. 10 mm	brownish
Trichloroethylene	50 ppm	5	approx. 10 mm	brownish green
Trichloroethylene	above liquid	5	completely coloured	faded yellow
Xylene	500 ppm	5	approx. 10 mm	brownish
Xylene	above liquid	5	approx. 10 mm	brownish

Some examples of substances which do not react are given in table 4 and table 5 lists some examples of substances which react without producing a colour stain.

TABLE 4

NO REACTION i.e. no colour indication is obtained on the Polytest tube from:

Carbon Dioxide	(CO_2)
Hydrogen	(H_2)
Methane	(CH_4)
Ethane	(C_2H_6)

These gases flow through the white filling preparation without any reactions; the preparation does not change its properties.

TABLE 5

The following substances react on the Polytest tube without colour indication:

Ammonia	(NH_3)
Chlorine	(Cl_2)
Hydrogen chloride	(HCl)
Nitric acid	(HNO_3)
Sulphur dioxide	(SO_2)
Nitrogen dioxide	(NO_2)

These gases react with the white preparation in the Polytest while flowing through but the filling retains its white colour unchanged.

Ethyl Acetate 200/a tube. The other tube which has a very broad spectrum response is the Ethyl Acetate 200/a tube. The reagent system in these tubes consists of a chromate salt and sulphuric acid, this combination is popularly known as chromosulphuric acid. It is known to react with a wide range of organic compounds and some examples are given in table 6.

TABLE 6
Examples of Qualitative Indication on the Ethyl Acetate 200/a Tube

Substance	Concentration	Number of strokes of the bellows pump	Length of discolouration	Notes on the indication
Acetone	3000 ppm	5	approx. 10 mm	greenish
Acetone	above liquid	5	completely coloured	greenish
Benzene	500 ppm	5	completely coloured	very pale grey
Benzene	above liquid	5	completely coloured	greenish grey
Ethyl Alcohol	2000 ppm	5	approx. 5 mm	greenish
Ethyl Alcohol	above liquid	5	approx. 20 mm	greenish

TABLE 6 (contd)

Substance	Concentration	Number of strokes of the bellows pump	Length of discolouration	Notes on the indication
Octane	100 ppm	5	approx. 10 mm	grey-brown-greenish
Octane	above liquid	5	completely coloured	greenish
Toluene	500 ppm	5	approx. 10 mm	greenish grey
Toluene	above liquid	5	completely coloured	greenish grey
Xylene	500 ppm	5	approx, 10 mm	greenish brown
Xylene	above liquid	5	completely coloured	greenish brown

Again, however, it must be realised that not all pollutants will react on this tube. Table 7 gives some examples of substances that do not react and table 8 lists examples of substances which react without producing a distinct colour change.

TABLE 7

No reaction i.e. no colour indication is obtained on the Ethyl Acetate tube from:

> Ethane
> Carbon Dioxide
> Carbon Monoxide
> Methane
> Hydrogen

These gases flow through the orange-coloured filling preparation without any reaction; the preparation does not change its properties.

TABLE 8

The following substances react on the Ethyl Acetate tube without indication (or with only a minor fading or intensification of the colour of the indicating layer):

Ammonia (NH_3)
Chlorine (Cl_2)
Hydrogen Chloride (HCl)
Nitric Acid (HNO_3)
Nitrogen Dioxide (NO_2)

When these gases are passed through the tube, they react with the orange-coloured preparation of the ethyl acetate tube; however, the colour of the indicating layer changes but little due to fading or intensifying of the colour.

Similar examples could be given of detector tubes that will show a broad spectrum response to halogenated hydrocarbons, to amines and basic substances or to acid vapours.

These examples show that even when the pollutant is a complex mixture, it is possible to obtain a qualitative indication of its presence in the atmosphere. This indication is frequently sufficient to provide a go/no-go indication for safety purposes.

SPECIFIC PROBLEMS

It is worth mentioning at this point, that if the details of a specific problem are given to the detector tube manufacturer, he is frequently able to advise on the best detector tube for use, to indicate an approximate calibration and in some circumstances can arrange special calibration tests.

FIELD CALIBRATIONS

Another approach to the problem of monitoring vapours from complex mixtures is to undertake a series of ad-hoc field trials. In this method a measurement is taken with a detector tube whilst at the same time an air sample is taken with an activated charcoal tube. Alternatively, the detector tube reading may be compared with the result from some suitable portable analytical instrument. The charcoal tubes are subsequently analysed by standard methods, N.I.O.S.H. (6) and the results compared with the detector tube readings. If a reasonable correlation is obtained, then future tests may be made with the detector tube alone. This will obviously be cheaper, quicker and simpler than continuing with sample collection and analysis. It may be necessary in the initial stages of these trials to investigate the performance of a number of different detector tubes to see which type gives the best stain under the test circumstances. It is advisable to periodically repeat the correlation tests to ensure that no changes in the conditions of measurement have occurred.

CHOICE OF TUBE FOR VARIOUS HYDROCARBONS

Since the number of detector tubes available for hydrocarbons is quite large, it is not always easy to determine the best tube to use, tables 9a, 9b and 9c give some suggestions of the likely best tube for a range of simple alkanes, alkenes and aromatic hydrocarbons.

TABLE 9a
DETECTOR TUBE TO USE FOR COMMON ALKENES

Formula	Tube to Use	Range
CH_4, C_2H_6	Natural Gas	Qualitative 0.5-1.3% C_3H_8
C_3H_8, C_4H_{10}	Hydrocarbons 0.1%/b	0.1-0.8% C_4H_{10}
C_5H_{12}	n-Pentane 100/a	100-150 ppm
C_6H_{14}	n-Hexane 100/a	100-3000 ppm
C_7H_{16}	Cyclohexane 100/a	100-1500 ppm
C_8H_{18}	Petroleum Hydrocarbons 100/a	100-2500 ppm C_8H_{18}
	Hydrocarbons Test	500-2500 ppm C_8H_{18}
C_9H_{20}, $C_{10}H_{22}$	Hydrocarbons 2	2-23 mg/l
$C_{11}H_{24}$, $C_{12}H_{26}$, $C_{16}H_{34}$	Petroleum Hydrocarbons for low concentrations	about 200-500 ppm

TABLE 9b

Formula	Tube to Use	Range
C_2H_4	---------- Ethylene 0.5/a or 50/a	0.5-10 ppm
	Hydrocarbons 0-1%/b	50-2500 ppm
C_3H_6 C_4H_8	Olefines 0.05%/a	1-55 mg/l

TABLE 9c

Formula	Tube to Use	Range
C_6H_6	Benzene 0.5/a	0.5 - 10 ppm
	Benzene 5/a	5 - 40 ppm
	Benzene 5/b	5 - 50 ppm
	Benzene 0.05	0.05- 1.4 mg/l
$C_6H_5(CH_3)$	Toluene 5/a	5 -400 ppm
	Toluene 25/a	0.1 - 7.0mg/l
$C_6H_4(CH_3)_2$	Toluene 25/a	0.1 - 7.0mg/l
$C_6H_3(C_2H_5)$	Ethylblenzene 30/a	30 -600 ppm
	Toluene 25/a	0.1 - 7.0mg/l
$C_6H_3(CH_3)_3$	Toluene 25/a	0.1 - 7.0mg/l
$C_6H_5CH(CH_3)_2$	Toluene 25/a	0.1 - 7.0mg/l

When faced with the problem of measuring the vapour concentration of a distillation faction, it is necessary to clarify whether the major constituents are alkanes, alkenes or aromatic hydrocarbons and the boiling range. Then it is possible to assume that a significant portion of the mixture consists of a single hydrocarbon with a boiling point in the middle of the quoted range. A suitable detector tube may be chosen for this single substance. The mixture will usually give a good indication on the chosen tube and the calibration is likely to be within $\pm$ 50%. This is frequently sufficient for many safety purposes. Such tests will certainly highlight problem areas which can then be the subject of a more detailed study.

CONCLUSION

When attempting to monitor the working environment for the vapours of common industrial solvents, it is often possible to devise a simple, quick, cheap test using a detector tube. This is the case even when the solvent used is a mixture of several components. The results of the qualitative or semi-qualitative tests are usually adequate for routine safety purposes. If such a routine test indicates an area with a high vapour concentration, then and only then is it necessary to employ more costly sophisticated techniques.

REFERENCES

1. Thain W (1980) *In* "Monitoring Toxic Gases in the Atmosphere for Hygiene and Pollution Control", Pergamon Press, Oxford.

2. Leichnitz K. (1981) "Prufrohrchen Messtechnik", Ecomed, Landsberg/lech, West Germany.

3. Walton J.A. *In* "Handbook of Occupational Hygiene", Vol. 1, Section 4.1.2., Kluwer, Brentford.

4. Guidance Note EH15/80 from the Health and Safety Executive "Threshold Limit Values for 1980", H.M.S.O., London 1981.

5. Leichnitz K. (1979) *In* "Detector Tube Handbook" 4th edition, p.86, Draeger Lubeck.

6. N.I.O.S.H. Manual of Analytical Methods, 2nd edition, Vol. 1, Section 127, U.S. Department of Health, Education and Welfare, Cincinnati.

THE USE OF AIR MONITORING BADGES
FOR HEALTH PROTECTION IN HANDLING
CHLORINATED SOLVENTS MIXTURES

F.H. van Mensch and J.W.A. de Graaf[*]

Akzo Zout Chemie Nederland B.V.

[*]*7554 RS Hengelo (O), Box 25, NL*

1. SURVEY

The subjects to be dealt with are
a) Applications of chlorinated solvents.
b) Toxicological properties.
c) Safety precautions, directives.
d) Concentration control at the workplace.
e) Accurate badges; new tool in personal monitoring.
f) Some results.

2. DETAILS

a) Applications

In the case of the products methylchloride, dichloromethane, chloroform, carbon tetrachloride, trichloroethylene perchloroethylene, 1,1,1-trichloroethane and ethylenedichloride the accent of some is, as you can see in table 1, on application as a raw material for subsequent products (e.g. CH_3Cl is used in the production of silicones, while $CHCl_3$, CCl_4 and C_2Cl_4 are used in the production of fluorocarbons); moreover, the accent is on the application for special purposes such as extractions (CH_2Cl_2, $CHCl_3$) and blowing agent (CH_2Cl_2).

TABLE 1

Chlorinated Solvents and their Applications

Compound	b.p. $^\circ$C	special	wide range
CH_3Cl	-24	silicones	
CH_2Cl_2	39.8	blowing agent extractions	paint strippers degreasing
$CHCl_3$	61.2	fluorocarbons	
CCl_4	76.4	fluorocarbons	
H_3CCCl_3	74.1		degreasing
ClH_2CCH_2Cl	84	vinylchloride	
$ClHCCCl_2$	87.1		treatment of textiles/metals
Cl_2CCCl_2	121.2	fluorocarbons	treatment of metals/textiles

Since the applications mentioned before pertained to closed systems, the safety in using these components has been guaranteed rather well.

On the other hand there is a great variety of large scale applications of CH_2Cl_2, Tri, Per and 1,1,1-Tri for a wide range of cleaning purposes, like dissolution, extraction, removal of oils and fats, including the application as a solvent in adhesives and in paint stripping.

These large scale applications will, as a rule, require much more safety care. As a matter of fact there is a wide range of locations with different qualities and quantities where cleaning is done whether or not by means of equipment not always operated optimally. This being the consequence of lack of efficient instructions-for-use and efficient control.

These products offer a very favourable combination of technical and physico-chemical properties as a result of which they are considered irreplaceable for a number of applications. We just mention :
boiling point range 40 $^\circ$C - 120 $^\circ$C,
incombustibility (exceptions CH_3Cl and EDC),
stability (with addition of stabilizers),
high dissolving power.

Therefore, these products are worldwide used on a scale of hundreds of thousands of tons, and it looks like that this consumption volume will remain in the same order of magnitude, although now and perhaps in the near future some decline may be perceived.

The estimated production volume for West Europe is given in table 2.

TABLE 2

Estimated production volume (West Europe, tons, 1980)

Dichloromethane	230.000
Chloroform	80.000
Carbon Tetrachloride	340.000
Trichloroethylene	220.000
Perchloroethylene	270.000
1,1,1-Trichloroethane	160.000

b) Toxicological properties

Absorbed through the lungs, stomach or skin the chlorinated solvents show a more or less violent influence on the central nervous system resulting in:
 narcosis-anaesthesia;
 unconsciousness;
 possible attack and damage of liver and kidney and sometimes of heart and brains. Carbon tetrachloride especially is a relatively virulent liver toxicant, working anaesthetically as well. The above is rather dependent on the intensity and time of exposure. Besides, individual sensitivity is very different. Some substances (CH_3Cl and EDC) show a delayed effect, while almost all also affect eyes and skin;
 nausea, inclination to vomit, dizziness and other phenomena of disarrangement of the digestion are symptoms, which often only appear after some time.
 The metabolites Carbon Monoxide (from CH_2Cl_2) and Trichloroethanol and Trichloroacetic Acid (from Tri) contribute to the poisoning also, as do the decomposition products phosgene and hydrochloric acid.
Cases of acute or sub-acute deadly poisoning (sometimes dozens of cases in about a century) by all chlorinated solvents are known. Initially many of these cases were due to technical defects of domestic cooling equipment (CH_3Cl as refrigerant) but there were also cases of guilty negligence or deliberate self-poisoning.

Growth of applications (e.g. of Tri, Per, 1,1,1-Tri and Dichloromethane) since the thirties, together with increased knowledge and understanding of the detrimental effect implied risks in professional handling of these solvents to decrease.

Although it could be expected that there might also be a clear relation between a deepening insight into the properties and the number of non-professional deadly casualties, this relation hasn't been found. Abuse of the solvents remained one of their causes.

c) Safety precautions, Directives

Safe production, transportation and processing require a series of measures all directed towards protection of the workers' health.
Control includes checking the reliability of procedures and apparatus, as well as of fulfilment of instructions. In this respect some of the many rules, instructions, memoranda and legal regulations can be mentioned.
In the UK e.g. :
Health and Safety at Work Act 1974,
The Code of Practice for the recovery of chlorinated solvents and other halogenated organic compounds (Edited by Chemical Recovery Association 1975),
Waste Management Paper no. 9 on halogenated hydrocarbon solvent wastes from cleaning processes (Technical Memorandum edited by the Department of Environment 1976).
Authorities in the F.R.G. have since 1977-1979 made :
Safety rules for cleaning equipment,
Safety rules for the use of recovery equipment,
the so called Loesemittelverordnung with Danger and Safety (R and S) sentences, as have been explained in full extent by several speakers on wednesday.
In the Netherlands a new Act on working conditions will be effective in 1982/1983. The essential qualities and new features embodied in this act appear from a number of articles defining the duties of the employee some of which are summarized below. The employee is obliged i.a. :
to exercise the necessary caution and good care in order to avoid dangers to health,
to properly use the personal protective auxiliary equipment and protections,
to make himself familiar with the directives in an active way,
to comply with the instructions, etc. etc.

And last but not least there is the European Common Market Directive on the Protection of workers from harmful exposure to chemical, physical and biological agents at work no. 80/1107/EEC of 27 Nov. 1980.

All instructions and directives aim at:
good maintenance of equipment,
appropriate air ventilation,
suitable vapor suctions, and
periodical health check by the medical service department.
In figure 1 the increase of the safety idea is illustrated with the help of a survey of TLV's.

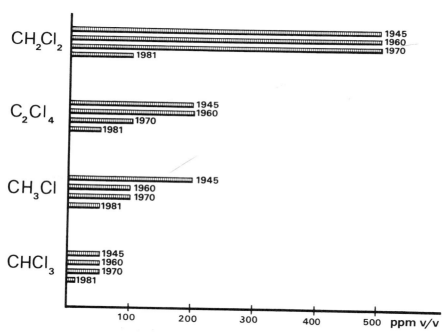

Fig. 1. *Survey of TLV's (period 1945-1981)*

d) *Concentration control at the workplace*

Maintenance, as mentioned before, periodical inspection and check of the atmosphere should take place near the plant unit and at the workplace, by means of continuous and discontinuous concentration measurements in the direct vicinity of equipment and in the atmosphere of the workshop.

Besides, personal check should be carried out by means of monitoring of vapour inhalation and by means of biological monitoring (urine, blood and liver function examination) This in accordance with the fondest wish of Mr. Burgess, expressed in his yesterday afternoon's lecture.

Apart from the available large and often very expensive equipment, badges are to be had from several suppliers (DuPont, 3M, Real, Abcor, and others).

To keep the atmosphere in the workshop under control, following equipment is i.a. available :
MIRAN : can determined one component at a time by adjusting at one certain IR-wavelength,
GASTEC : is not explosion proof, measures total carbon,
DRAEGER and AUER (MSA) tubes for spot tests, which can be read directly by color reaction. Accuracy is sometimes questionable.

For checking the employee, following possibilities exist: preconcentration adsorption tubes at breathing level, connected with sampling pumps. Weight is varying from about 300 grammes to about over 900 grammes. The equipment is to be carried at waist level; Desorption of tube content and analysis by means of Gas Liquid Chromatography are a must. Some suppliers are Sipin, Draeger, DuPont, MSA, Cape Industries.
Badges we mentioned before. As a matter of fact the same holds for them; desorption and GLC-analysis are necessary (1-6).

e) Accurate badges; new tool in personal monitoring

Handling of too voluminous equipment to be carried on the man might be difficult.
For this reason it was suggested to use a lightweight adsorption badge fixed at breathing level. However the problem with badges is to let simplicity and sufficient accuracy go hand in hand.

We have experienced that a simple badge can adsorb mixtures of chlorinated solvents and by experimental verification through static and dynamic calibration we have proven its reliability.
The badges tested by us have an accuracy of approximately 5 % or less, and it appeared that the individual compounds don't influence each other in their adsorption behaviour. The badge we chose for our experiments (Fig. 2) consists of a round aluminium little cup of 5 cm diameter and a thickness of about 0.8 cm, provided with a membrane with approximately 1 g activated carbon placed in it.

Weight of the badge is 35 g.

Under almost all circumstances sufficient venting of air takes place, and so the concentration gradients of the compounds to be determined are maintained over the membrane.

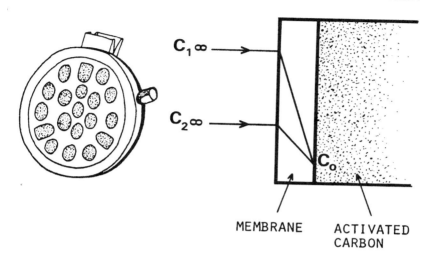

Fig. 2. *Badge and Slope of concentration gradients*

Slope of these gradients is only dependent on the concentration of the components in the ambient air.
The average ambient air concentration can then be determined after analysis of the quantity of adsorbed component in relation to the time of exposure. To this end a calibration factor has to be determined beforehand, indicating the relation between the ambient concentration and the quantity of matter adsorbed after some time. Such calibration factor is dependent on the properties of the membrane and those of the activated carbon.
So, by means of :
known calibration factor k,
known time of exposure t,
measured weight of adsorbed component w,
we find the concentration in the ambient air, c, to be

$$c = \frac{kw}{t}$$

The rather simple summary of the mathematical derivation of k is as follows :

where N = diffusion velocity (moles/sec)
    D = diffusion coefficient (cm^2/sec)
    A = membrane surface
    x = distance from the front of the diffusion layer
    C = concentration of the component at x
    C_∞ = concentration in the ambient air
    λ = diffusion layer thickness

now (Fig. 3) :

$$N = - DA\frac{dc}{dx}$$

after integration over λ :

$$N = DA\frac{C_\infty}{\lambda}$$

after time t total quantity of adsorbed component is :

$$w = Nt = DA \cdot \frac{C_\infty t}{\lambda} \; ; \; (\frac{\lambda}{DA} = k),$$

so : $w = \frac{C_\infty t}{k}$,

and consequently

$$C_\infty = \frac{kw}{t}$$

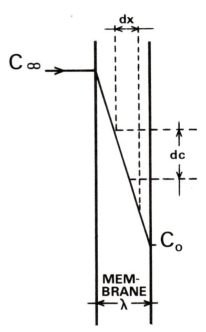

Fig. 3. *Concentration gradient in the membrane*

f) *Results*

In conclusion, here are - by way of figures - some results allowing an impression of the daily practice. Just for the sake of illustration, here first is an example of a set of calibration factors of one badge :

$k_{CH_2Cl_2} = 0,33$ $k_{CHCl_3} = 0,25$

$k_{C_2Cl_4} = 0,29$ $k_{EDC} = 0,19$,

while finally in Table 3 examples are given of concentrations found in
laboratory,
pilot plant, and
production site.

TABLE 3

Concentration in laboratory, pilot plant and production site (ppm, v/v)

location \ component	CH_2Cl_2	$CHCl_3$	CCl_4	C_2Cl_4
Laboratory and pilot plant	1.1-4.3	<0.1-1.1	-	<0.1-0.3
Production site, 124 observations (Oct. 1981- Febr. 1982)	<0.1-8.9 one extreme value: 210	<0.1-8.2 one extreme value: 113	<0.1-1.9 three extreme values: 2.1, 9.8 and 26 resp.	<0.1-6.7 two extreme values: 30 and 71 resp.

3. CONCLUSION

Results are so reliable that plant management has decided to introduce the system of personal monitoring on a more or less permanent basis.

4. REFERENCES

1. West, P.W. and Reiszner, K.D. (1976). *Proc. Specialty Conf. on toxic substances in the air env.*, Nov. 7-9, 1976, (Ed. by Air Pollution Control Ass. Pittsburgh, 1977).
2. West, P.W. and Reiszner, K.D. (1978) *Am Ind. Hyg. Ass. J.* 39, 645-650.
3. Holliday, M.M. and Anderson, J. (1980) *Analyst* 185, 289-292.
4. Hickey, J.L.S. and Bishop, C.C. (1981) *Am.Ind. Hyg. Ass. J.* 42, 264-267.

5. Stricoff, R.S. and Summers, C. (1981)
   Am. Chem. Soc. Symp. Series 149, 209-221.
6. Lautenberger, W.J. et al. (1981)
   Am. Chem. Soc. Symp. Series 149, 575-586.

LUNG UPTAKE OF ISOPROPANOL IN INDUSTRIAL WORKERS

F. Brugnone, L. Perbellini, P. Apostoli

*Istituto di Medicina del Lavoro, Universita
di Padova Policlinico di Borgo Roma
37134 Verona, Italy*

1. INTRODUCTION

Isopropanol is a solvent available for a large number of uses both in industry and in the home. In the past it was used for tepid sponging in feverish children (1). It is well known to be responsible for potentiation of hepatic and renal toxicity caused by haloalkanes (2-5). It is also suspected to be a carcinogenic agent after the discovery of an abnormal incidence of paranasal and sinus cancers in employees involved in isopropyl alcohol manufacture (6). In spite of its widespread use and its definitely harmful effects in humans, isopropanol uptake and metabolism have never been studied thoroughly in man.

2. MATERIALS AND METHODS

We studied environmental exposure to isopropanol in a group of workers employed in a printing works during a normal afternoon work shift. The investigation was carried out by examining environmental air, alveolar air, venous blood and urine.

Environmental and alveolar air, which were instantaneous and collected simultaneously, were sampled hourly. Venous blood was sampled at two hour intervals. Urine samples were collected before exposure after the end of work and then the next morning. A gas chromatographic technique was used for the analysis of air, blood and urine. The head space technique was used for blood and urine. Alveolar ventilation was measured during the work shift.

3. RESULTS

Table 1, which very briefly summarizes our findings is meant to emphasize that with a mean isopropanol exposure of 220 mg/m^3 (range 8-647 mg/m^3), we were in all cases able to find isopropanol in alveolar air but unable to find isopropanol either in blood or in urine. On the other hand, without our detecting any acetone in the atmosphere, we did find acetone in alveolar air, blood and urine with a concentration which was definitely higher than can be found under normal physiological conditions.

TABLE 1 Mean concentrations of isopropanol and acetone in environmental air, alveolar air, blood and urine.

			ISOPROPANOL			ACETONE		
		No	$\overline{M}$	SD	RANGE	$\overline{M}$	SD	RANGE
Environ. air	(µg/l)	90	220	187	8-647	0	0	
Alveolar air	(µg/l)	90	113	155	3-439	23	20	3-93
Blood	(µg/l)	42	0	0		4.5	3.1	0.8-15.6
Urine	(µg/l)	12	0	0		5.0	5.5	0.9-18.2

Figure 1 indicates that the mean alveolar concentration of isopropanol (shown in the 2nd line) is correlated with the mean environmental concentration os isopropanol (shown in the upper line). Because of this the variations on the two isopropanol concentrations, during the 7 hours of the work shift, follow a parallel course. In the same figure, the mean alveolar concentration of acetone (shown in the lower line) is correlated with the mean acetone concentration in blood (shown in the 3rd line). As can be seen, alveolar and blood acetone concentrations rise during the first 3 hours of exposure. Following this time, the variations in alveolar and blood acetone concentrations are, on the whole, parallel with the variations in isopropanol in environmental and alveolar air.

Statistical analysis of the data of alveolar (Ca) and environmental (Ci) isopropanol concentrations showed that a highly significant correlation existed between Ca and Ci (Ca=0.539 Ci - 16; r=0.9186; n=90; p<0.001).

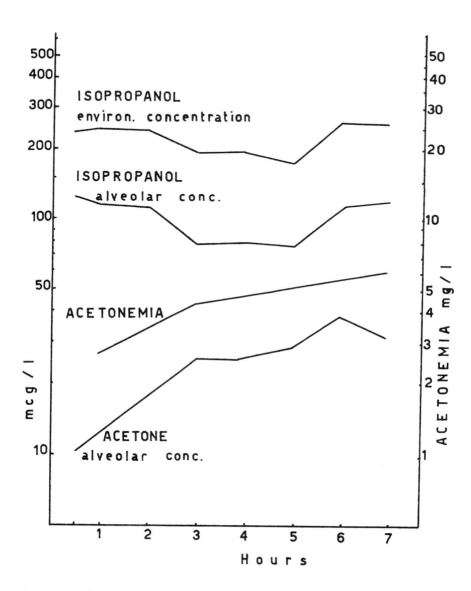

Fig. 1 Variations of environmental and alveolar isopropanol concentrations and alveolar and blood acetone concentrations during the work shift.

As shown in Table 1, we detected acetone in alveolar air, in blood and in urine but not in environmental air. Acetone is a physiological metabolite which can be found normally in human biological media. Its mean concentration before isopropanol exposure was 3.7 µg/l (S.D.=3.8), 1.4 mg/l (S.D.=0.5) and 3.9 mg/l (S.D.=7.6) in alveolar air, blood and urine respectively. The concentration of acetone following exposure to isopropanol, as shown by the data reported in Table 1, was definitely higher than before exposure both in alveolar air, blood and urine. The examination of individual data collected before the during the work shift showed that a highly significant correlation existed between the blood (C_b) and the alveolar (C_a) concentrations of acetone (C_b-101 C_a + 1.8; r=0.6728; n = 54; p<0.001).

The concentration of acetone, both in alveolar air and in blood, was correlated with the alveolar concentration of isopropanol, but the ratio between acetone and isopropanol given by the slope of the regression line varied with the length of exposure. In particular the alveolar acetone concentration corresponded to about 8, 12 and 18% of alveolar isopropanol concentration at the 1st, 2nd and 3rd hours respectively of the work shift. On the other hand blood acetone concentration corresponded to 9 and 30 times the alveolar isopropanol concentration at the 1st and 3rd hours respectively of the work shift.

Table 2 compares alveolar isopropanol uptake with the acetone eliminated during the work shift by the lungs and in urine. As can be seen the amount of acetone eliminated by the lungs expressed as a percentage fraction of absorbed isopropanol ranged between 11 and 40% while the amount of acetone eliminated in urine during exposure was between 0.04 and 0.73% of the isopropanol absorbed in the same time.

4. COMMENTS

It seems to us that 3 aspects of our findings are worth emphasizing. Firstly there is the very good correlation between the alveolar and environmental concentrations of isopropanol. The assessment of this correlation suggests that at any time (and on the basis of the C_a/C_i ratio) we can know the individual exposure by testing isopropanol concentration in the alveolar air of the workers.

The second aspect is that we cannot use the blood isopropanol concentration as an index of exposure when the

TABLE 2 Comparison between adsorbed isopropanol and eliminated acetone during the work shift.

NAME	ISOPROPANOL ALVEOLAR UPTAKE mg	ACETONE ELIMINATED DURING EXPOSURE TO ISOPROPANOL	
		by lungs mg	in urine mg
1)DM	1680	180 (10.7%)	2.5 (0.14%)
2)VE	1805	273 (15.1%)	8.1 (0.45%)
3)ZA	1410	173 (12.3%)	9.6 (0.68%)
4)NI	232	84 (36.2%)	1.7 (0.73%)
5)RI	373	90 (24.1%)	1.2 (0.32%)
6)RV	434	110 (25.3%)	2.0 (0.46%)
7)BE	492	75 (15.2%)	0.4 (0.08%)
8)BO	443	125 (28.2%)	0.2 (0.04%)
9)PA	688	274 (39.8%)	0.6 (0.09%)

Alveolar isopropanol uptake: $C_a \times R/(C_a/C_i) \times \dot{V}_A \times T$ with
C_a = alveolar conc. ($\mu g/l$), R = alveolar retention $(1-C_a/C_i)$,
C_i = environmental conc., $\dot{V}_A$ = alveolar ventilation, T = time of exposure.
Alveolar acetone elimination: $C_a \times \dot{V}_A \times T$.
In brackets: eliminated acetone expressed as a fraction of alveolar isopropanol uptake.

environmental concentration of isopropanol found by us was 647 mg/m^3 (Table 1) and we never detected isopropanol in workers' blood. In our opinion this can be explained by bearing in mind that the apparent volume of isopropanol distribution in the human body is large enough to determine in relationship with the environmental isopropanol concentrations found by us, a blood concentration lower than our technical limit of detection which was 1 mg/l.

The third aspect deals with the production and the excretion of acetone in relationship with isopropanol exposure. It has been known for a long time that acetone is a metabolite of isopropanol (7) but it has never been studied in humans exposed to isopropanol. Folland et al. (5) in workers occupationally exposed to isopropanol calculated that the blood acetone concentration should have been 3 to 30 times more than the normal range. In our research we found that the alveolar acetone concentration after 3 hours of exposure was, on average, 18% of the alveolar isopropanol concentration. Moreover we found that the mean acetonemia was correlated with the alveolar isopropanol uptake with a slope suggesting that an acetonemia of 3.5 mg/l corresponded to an isopropanol uptake of 1 g.

In conclusion our experimental data allow a deep insight into the occupational exposure to isopropanol. On the basis of this it is certain that the biological monitoring of isopropanol in alveolar air and of acetone both in alveolar air and in blood could lead to a more safe use of isopropanol in industrial plants.

REFERENCES

1. Garrison R.F.,(1953). J.A.M.A., 152,317-8.
2. Cornish H.H., Adefuin J.,(1967). Arch. Environ. Health, 14, 447-449.
3. Traiger G.J.,Plaa,G.L., (1971) Toxicol. Appl. Pharmacol. 20, 105-112.
4. Traiger, G.J., Plaa, G.L. (1974) Arch. Environ. Health, 28, 276-278.
5. Folland, D.S., Schaffner, W., Ginn, H.E. Crofford, O.B. McMurray, D.R. (1976) J.A.M.A., 236, 1853-1856.
6. NIOSH (1976) Criteria for a recommended standard. Occupational exposure to Isopropyl Alcohol HEW Publication N. 76-142.
7. Kemal, H., (1927) Biochem. Z., 187, 461-466.

THE CONTROL OF INDUSTRIAL EXPOSURE TO SOLVENTS

M.K.B. Molyneux

Shell (U.K.) Limited
Occupational Hygiene Unit
Carrington, Urmston, Manchester M31 4AJ

For the purpose of this presentation it is helpful to centre attention on 'the controller', a fictitious individual, having the necessary intellect, executive authority and resources with which to achieve the single objective, which is to control industrial exposure to solvents. One recognises in the controller the essential medical, hygiene, toxicological, engineering, technological, operational, managerial and political skills which are required to undertake the task.

On first taking the position in the period 1955 - 1960 the controller would have had access to authoritative information on the possible harmful effects of industrial exposure to solvents in the Report of the Medical Research Council as published by Browning (1). Combining this information with contemporary views on homeostasis and the TLVs of the time, the controller would set out to reduce the acute and chronic harmful effects of solvents to an acceptable level. Knowledge of these effects was largely based on human experience, and the relationships between dose, effect, irreversible change and recovery were reasonably well documented. The principal target organs were CNS, liver, kidney, blood and skin.

The position has changed. New possible harmful effects have been identified and hygiene standards have been driven down to more stringent levels. Of the ten classes of solvents identified by Browning (1), shown in Table 1, seven now have individual solvents listed as substances

TABLE 1

GROUPS OF SOLVENTS QUOTED BY BROWNING (1953)

- * HYDROCARBONS
- * CHLORINATED HYDROCARBONS
- ALCOHOLS
- * ETHERS
- ESTERS
- KETONES
- * GLYCOLS
- * AMINES
- * NITRO COMPOUNDS
- * MISCELLANEOUS

* Individual solvents listed in Appendix A2 of ACGIH 1981 - Industrial Substances Suspect of Carcinogenic Potential for MAN.

suspect of carcinogenic potential for man by the American Conference of Governmental Industrial Hygienists (2). The controller now has an extended list of possible harmful effects which include peripheral neuropathy, carcinogenicity, mutagenicity, reproductive effects and kidney malfunction. Each of these effects have their own dose response, recovery characteristics and latent periods, about which the controller lacks essential definitive information.

Examples of changes in hygiene standards can be seen by comparing the values quoted by Luxon (3), at the conference of the British Occupational Hygiene Society on Organic Solvents in 1966, with the ACGIH TLVs for 1981. Of thirteen solvents listed, seven now have lower values (Table 2). Two other solvents, Stoddard Solvent and n-Hexane also have greatly reduced TLVs.

The position has also changed in another important aspect, that of feedback. For these newly recognised possible harmful effects the controller has little or no biological feedback on the extent to which his actions are being effective. In the worst case the amount of effort put into control may bear no relationship to the real hazard or to the benefit achieved. Current workplace statistics give little guidance because, where they do exist, they tend to relate to overt effects of poisoning or mortality, with complications of latent period and uncertain historical exposure data. Where statistical data do exist there are questions of method and quality, which cast doubt on the findings. Similar questions have been raised over toxicological studies which form the principal alternative source of information. The controller is forced to accept long delays between the controlling action and feedback, if any, on the benefit, or the negative effect, of the action.

The increased awareness of possible carcinogenic and mutagenic effects has resulted in a harder attitude by exposed persons and more strict government policies. In the U.K. this is shown by the 'no carcinogen' proposals by trade unions and the Precautionary Policy of the Health and Safety Executive (4). Here the move is towards the exercise of rigorous control measures for substances whether or not they are known to have specific harmful effects.

TABLE 2

CHANGES IN HYGIENE STANDARDS OF SOLVENTS

	1966	1981	
Carbon tetrachloride	10	5	ppm
Carbon disulphide	20	10	
Benzene	25	10	
Cyclohexanone	50	25	
Methyl isobutyl ketone	100	50	
Trichloroethylene	100	100	
Methyl alcohol	200	200	
Methyl ethyl ketone	200	200	
Toluene	200	100	
Xylene	100	100	
1.1.1 Trichloroethane	350	350	
Cyclohexane	400	300	
Ethyl acetate	400	400	

Based on Luxon (1966) and TLVs (1981).

The minimum requirement is to keep exposures below the exposure limit and then, depending on the substance, there may be a further requirement to reduce exposure as far as reasonably practicable. Consequently the controller may need to adopt more than one standard for a harmful substance depending on opinion of dose response and its possible harmful effects.

This diversity of potential harmful effects, the stricter exposure values, the deficiencies in toxicological data, the lack of feedback, the hardening of attitudes, the stricter policies, have all had an impact on the control of industrial exposure to solvents and, compared with the previous situation, have made the task of the controller more difficult and less predictable.

Fortunately three aspects of control remain unchanged and are as valid now as they were in the period 1955-60. First, for most substances the harmful effect is expected to be directly related to dose, which is a function of amount and time. Second, there are only three routes of intake for solvents, inhalation, skin, and to a lesser degree, ingestion. Third, the principles of control remain valid, which are listed in Table 3. Of these, emphasis is placed on the correct recognition of the route(s) of intake, which is the foundation of all subsequent actions. It then remains to apply the principles of control more effectively in order to meet the stricter standards. The controller would have two thoughts in mind when applying the principles in Table 3; first, that reading from top to bottom, the capital cost decreases, the effort to maintain the method increases and the safety factor decreases; second, that a system approach should be adopted based on cost and benefit, knowing that no single principle in isolation is likely to provide a satisfactory degree of control. Hence the need to combine technical, scientific, administrative, operational and medical expertise.

From experience the controller would note significant improvements in five of the ten principles, i.e. Segregation, Enclosure, Information, Education, Skin Protection.

Guidance on control by design using the principles of Segregation and Enclosure on petrochemical plant can be found in Brief and Lynch (5). Their approach for new plant is, initially, to obtain a relative risk of over-exposure, based on the inventory of materials, physical agents, exposure limits, and the probability of release. Given this information attention is then directed at the principal sources of release, which are listed as:- Sampling-5, Pump Shaft Seals-4, Flanges, Tankage, Filter Change-5, Liquid Waste Streams-4, Vents-6. The figures give the number of options which are quoted for minimising the releases. A similar systematic approach is recommended for the control of exposure from abnormal operations such as plant refits.

The same authors (6) emphasise the need to introduce design concepts at the beginning of new processes and discuss the types of input at the different stages of the project i.e:

>Process Development
>Preliminary Project Stage
>Design Specification Stage
>Detailed Engineering Stage
>Operating Manual
>Construction
>Start-up Stage

Laboratory design can be approached in a similar, organised way, taking as an extreme example, the control of carcinogens as prescribed by OSHA, as interpreted by Rappaport and Campbell (7). Here, the inventory separates prescribed materials from exempt materials (e.g. solutions having concentrations of less than a prescribed amount of carcinogen) and the control procedures can be considered under the following main headings:-

>Segregation - the identification of prescribed areas, amenities, glove box criteria, waste, effluent treatment, storage.

>Ventilation - consideration of critical factors such as recirculation, independent exhaust system, make-up air, air filtration, location of fan and filter, location of exhaust to outside air, filter efficiency, fume hood criteria, maintenance, decontamination.

Administration   – access control.

Materials   – working surfaces

Services   – isolated water supply, scrubbed vacuum lines.

Emergency   – accidental release, decontamination.

As a matter of principle these and other factors are likely to be considered in the design of any new laboratory installation but the extent to which they are introduced will depend on the severity of the potential hazard.

The controller would be aware of considerable progress in Information, its collection, storage and distribution. Examples in U.K. legislation are the Packaging and Labelling of Dangerous Substances Regulations (8) and the associated Notes of Guidance. These implement the corresponding EEC Directive 67/548/EEC. Through these and similar instruments the controller would have immediate access to basic information on the potential hazard and the approach to control. Further information is now distributed by manufacturers and suppliers in the form of health and safety data sheets from which the controller would take the first step towards formulating detailed control procedures.

Improvements in Education and Training assist the controller in two important ways, with the recruitment of trained personnel and the training of existing staff. In the U.K. there is now a structured professional training programme for hygienists operated through the British Examining and Registration Board in Occupational Hygiene, plus academic courses at graduate and post graduate level. These routes provide the feedstock to future specialist manpower in the occupational hygiene aspects of control. The training of occupational physicians has long been established in the U.K., and industry and unions have placed considerable effort into the education and training of managers, safety officers, safety representatives and workers. Hence the controller can now be supported by a more professional, more educated, more trained and more organised industrial population than previously.

One aspect of personal protection which apparently shows signs of marked improvement is Skin Protection. This can be accounted for in a number of ways, amongst which are better products, low cost, better availability, improved standards of personal hygiene and the fact that the skin is probably the easiest route of intake to protect. However, as with all methods of personal protection, the controller would be aware that low capital cost has to be weighed against the operational snags. Although there are several alternative materials e.g. polyvinyl chloride, polyethylene, nitrile rubber, butyl rubber and neoprene, there are insufficient data on the various characteristics to enable a quantitative assessment of performance to be made. Whereas, in many situations, they may provide a satisfactory safety factor for protection against solvents which are irritant, defatting, against systemic poisons, carcinogens and others, the operational snags remain and can be summarised as follows:-

o Solvent identification - particularly mixtures, individual components of which have different physical and chemical properties.

o Permeation - lack of information on the rate of permeation and the accumulation of contamination in contact with the skin.

o Perforation - degree of quality control on new items and the detection of perforation during use.

o Wearer acceptance - the deciding factor on when and where the protection is used.

o Durability - determining the "safe" life of the protection.

o Decontamination - the control of exposure arising from repeated use.

There is potential for improvement in two of the listed principles of control i.e. Work Method, and Ventilation. Control by work method is dependent on job design, awareness of the potential hazard, and know-how. In manual operations with solvents, work method can be designed to minimise exposure by detailed consideration of posture, method of application or transfer, relative positions of sources of release, quantities in proximity to the worker, rate of production, exposure time, and non-exposure time. Hence, work method design, which is effective at the interface between the operator and the potential hazard, provides an opportunity to control the two primary factors which regulate the intake i.e. concentration (or quantity) and time. Using this method the controller would expect to have most impact on jobs involving, for example, cleaners, adhesives, paints and laboratory bench work.

The laboratory fume hood can be used to illustrate the move towards achieving a satisfactory degree of containment, by enclosure and exhaust ventilation, at minimal cost. Standard practice would normally lead the controller to assess fume hood performance on the basis of air flow. The British Occupational Hygiene Society (9), for example, give recommended linear flow rates for general purpose use at 0.5 m/s and for rigorous control at 0.76 m/s. Recently, consideration has been given to assessing containment in relation to the exposure limit (e.g. TLV), taking into account the rate of emission of the solvent, the average face velocity, a factor of safety, and the area of the opening. For assessment purposes the required operational containment level can be compared with the fume hood containment level as measured or specified by the manufacturer. For many substances the required degree of containment may be achieved at lower air flow rates than those normally quoted, which may be a significant cost factor where a large number of hoods are being used. In comparing the advantages and disadvantages of the two approaches the controller would consider the effect of reduced air flow on performance in general and weigh this against the benefit in cost.

There are many industrial exposures to solvents which can not be controlled as neatly as by the laboratory hood. Extreme examples are the painting of ships, both external and internal, painting of engineering plant, and vehicles. Taking the case of brush painting inside ships' compartments, the main contributors to risk from solvent exposure are the remoteness from services, the confined space, the closeness of the worker to the emission, the large surface area of the emission, and low natural clean air supply. Given a range of saturated vapour concentrations of say, 4000 to 117000 ppm for the possible solvents used, it can be seen that high exposures may readily occur. Also given that the amount of clean air to dilute one pint of solvent to its TLV may typically range from 5000 to 500,000 ft^3 it can be seen that the rate of supply to achieve acceptable breathing zone concentrations can be technically difficult to achieve and to maintain. In practice the controller would recognise the need for high grade respiratory and skin protection, both of which can be applied in industry according to need.

Respiratory Protection is still a secondary option for the control of exposure to solvents but nevertheless can be highly effective. On paper the performance of the various types appears to be satisfactory, with nominal protection factors ranging from 7 to 2000 depending on type (British Standards Institution (10)). But, as for skin protection, the controller would be aware of snags, which can be summarised as follows:-

- o Face seal leakage - critical for filter type and depending on facial shape, size, activity, and adjustment.

- o Valve leakage - most critical for filter type, caused by distortion or contamination.

- o Maintenance - need for regular inspection, cleaning, replacement.

- o Duration - limit to the length of time worn, shortest for filter type but restrictive for all types.

- o Work rate - the impact on efficiency and limit to length of time worn.

- o Fitness - a limiting factor in the use of some equipment.

- o Filter Life - the lack of indication of residual life during use.

- o Air quality - the need to meet minimum criteria.

o Air quantity - need to meet minimum criteria.

In assessing the relative merits of the different control methods the controller would be aware of the continuing questions which are associated with skin and respiratory protection, weighing up the technical performance, the reliance on the human factor, the need for critical supervision, and the organisational effort to maintain these methods of minimising exposure.

Two control methods listed in Table 3 remain to be discussed, the first, Substitution, the last, Monitoring. Substitution has been a powerful tool for the controller, good examples being the move from benzene to toluene, the move from carbon tetrachloride to trichloroethylene, then to 1,1,1 trichloroethane. However, the controller may approach future prospects of substitution with circumspection in the knowledge that the spectrum of potential harmful effects by an individual substance may in reality be extensive and may not be obviously related to structure. The outcome may be that similar control procedures may need to be applied to a wide range of materials irrespective of their initial or ultimate harmful effects.

Monitoring provides the feedback on the effectiveness of the control system as it applies to one or other set of hazards. The methodology and the techniques for workroom air sampling, including diffusion samplers, have been discussed in previous papers, and indicate that the controller now has access to effective tools for this purpose. Together with biological monitoring (e.g. phenol in urine, trichloroacetic acid, mandelic acid) and medical monitoring, they form the essentials of the feedback system, and can be applied to the individual or used for epidemiological purposes. It may be that valid biological monitoring is potentially the most effective means of overcoming the deficiency in feedback.

It is reasonable to ask if progress on the control of industrial exposure to solvents has been made since the conference held by the British Occupational Hygiene Society in 1966. There has been no organised feedback to answer that question but in the areas highlighted in this paper the answer is probably affirmative. However, a

TABLE 3

PRINCIPLES OF CONTROL

SUBSTITUTION

SEGREGATION

ENCLOSURE

EXHAUST VENTILATION

DILUTION VENTILATION

WORK METHOD

PERSONAL PROTECTION

INFORMATION

EDUCATION

MONITORING

distinction may need to be drawn between the larger, highly organised manufacturers and suppliers, and the smaller user. It was with the small user in mind, in the U.K., that a number of occupational health and hygiene services were formed in the 1960s, some associated with Universities. The majority are still in existence but some failed. If small users of solvents still have unacceptable risks it raises the questions why and to what degree ? Is it due to inadequate labelling, lack of medical and toxicological data, ignorance of legal requirements and exposure limits, lack of resources for monitoring, lack of trained manpower, resistance to the introduction of control methods, lack of feedback, or is it cost ? Given the answers to these questions the controller would be better placed to apply the appropriate effort.

In retrospect, the concept of the controller, as a self contained individual serves to emphasise the degree to which the responsibility for control is delegated to many disciplines and many individuals. The better the knowledge, the training, the know-how and the organisation, the more likely that control of industrial exposure to solvents will be achieved. Also the closer to the workplace that these factors can be brought to bear, the more likely that control will be achieved.

REFERENCES

1. Browning E. (1953) Toxicity of Industrial Organic Solvents. HMSO. London.

2. American Conference of Governmental Industrial Hygienists. Threshold Limit Values for chemical substances in workroom air (1981). P.O. Box 1937, Cincinnati, OH 45201.

3. Luxon S.G. (1966) Annals of Occupational Hygiene **9.** No.4. 231-234.

4. Health & Safety Executive (1978) Toxic Substances - A Precautionary Policy. Environmental Hygiene 18.

5. Brief R.S., Lynch J. (1978) American Industrial Hygiene Association Journal. **39.** 620-625.

6. (1980) Ibid **4.** 832-835.

7. Rappaport S.M. and Campbell E.E. (1976) American Industrial Hygiene Association Journal. **37.** 690-696.

8. The Packaging and Labelling of Dangerous Substances Regulations (1978) Statutory Instrument 1978 No.209 HMSO London.

9. British Occupational Hygiene Society. A guide to the Design and Installation of Laboratory Fume Cupboards. (1975).

    Annals of Occupational Hygiene **18.** 273-291.

10. British Standards Institution. Recommendations for the selection, use and maintenance of respiratory protective equipment. BS.4275.1974.

HALOGENATED SOLVENTS IN INDUSTRY
CONTROL OF SOLVENT EXPOSURES

B.P. Whim

*Imperial Chemical Industries PLC Mond Division
P.O. Box 19, Runcorn, Cheshire, WA7 4LW*

1. INTRODUCTION

Halogenated hydrocarbons find widespread use in industry either as a raw material for other chemicals or as a solvent. A limited number are used as solvents and within that group a major division exists between the use as cleaning agents and as carriers for resins/elastomers in formulated products such as adhesives, printing inks and paints. This paper is concerned with the cleaning applications.

The four major solvents are trichloroethylene, tetrachloroethylene, (commonly known as perchloroethylene) 1,1,1-trichloroethane and 1,1,2-trichloro 1,2,2-trifluoroethane (known as FC113). Specific features of these compounds govern their use and Table 1 summarises the physico chemical properties of key importance.

Table 1
Halogenated Solvents: Properties, Uses

| Solvent | Flammability | TLV-TWA* ppm | Solvent Power | | Major Uses |
			Solubility Parameter	Kauri Butanol No.	
Trichloroethylene	NON FLAMMABLE	100	9.3	132	Vapour Cleaning
Perchloroethylene		100	9.4	90	Dry Cleaning
1.1.1-Trichloroethane		350	8.6	124	Cold Cleaning and Vapour Cleaning
1.1.2-Trichloro 1.2.2-Trifluoroethane (F 113)		1000	7.2	30	Precision Industry Cleaning and Dry Cleaning

* See U.K. Health and Safety Executive Guidance Note.

Since all four are non flammable other features assume greater importance in defining the area of application.

2 SOLVENT CHOICE

2.1 Metal Cleaning in the Engineering Industries

In the production cleaning of metal articles the solvent clearly needs to have a high capacity to dissolve oils, greases, and waxes and to be able to remove a wide variety of contaminants such as cutting oils, pressing aids and lubricants. The solvent properties of all are adequate for this duty but trichloroethylene and 1,1,1-trichloroethane predominate since in addition they have the right balance of volatility, surface tension and cost to meet the needs of industry.

In those industries where intermittent or short production runs are common the normal methods of cold cleaning by dipping, spraying and hand wiping are widely used since little or no capital is involved. These methods demand that the solvent used should be of low toxicity since emissions to the workroom are more likely than in other methods and for this reason 1,1,1-trichloroethane is commonly used.

Many industrial production sequences however require continuous cleaning methods and vapour cleaning in a variety of forms is standard practice. Trichloroethylene and more recently 1,1,1-trichloroethane are the major solvents since their physico chemical properties of non flammability, high vapour density and low latent heat are ideally suited to this application.

2.2 Electronics Industry

The electronics industry and particularly that sector involved with the production of printed circuit boards is typical of industry where critical cleaning to a high standard is essential and where the items involved comprise plastics, rubbers, resins and paint markings. The removal of contaminants such as flux residues and fingerprints etc demand selective solvent power such that cleaning without damage to the substrate occurs.

Trichlorotrifluoroethane (FC113) used in a wide variety of solvent compositions is the major solvent in these industries. Its low toxicity is also valuable in this application but its high cost precludes its more general use.

2.3 Drycleaning

The advent in the 1950s of triacetate fibres caused the replacement of trichloroethylene by tetrachloroethylene in drycleaning machines due to the latters milder solvent effect. Its use in unit shops is widespread in all developed markets.

Trichlorotrifluoroethane (FC113) is also used in drycleaning where its mild solvent action can be employed to considerable advantage in producing a high standard of cleaning.

3 METHODS OF USE AND OPERATING PRACTICES

3.1 Cold Cleaning

The scale of use here varies from infrequent hand wiping in maintenance schedules to large scale dipping or spray applications. Each application requires individual attention but certain key principles can be applied.

General Precautions

1   Good ventilation is important and, if possible local exhaust ventilation with venting away from the user. The main objective is to ensure that exposure to the solvent vapour is as low as possible and certainly below the TLV-TWA. Unnecessary breathing of solvent vapour should be avoided.

2   Entry into cleaning plants, pits, ducts or enclosed vessels containing solvent vapour must follow well known safety practices.

3   If there is a risk of splashing protective goggles should be worn.

4   Contact with the hands should be avoided as far as possible, but if occasional contact is likely PVC gloves should be worn to prevent removal of the natural grease from the skin.

5   Solvent contact with naked flames, red-hot surfaces, welding arcs or hot elements of electric heaters should be avoided and smoking discouraged.

6   Good working practices are important for safe and economic use. Cleaning methods should be such as to minimise solvent vapour emissions.

Good Housekeeping

1   Solvent should not be stored or carried in buckets or other open containers and its issue should be closely controlled.

2   Any containers should be clearly labelled.

3   In the event of a major spillage the area must be evacuated and ventilated thoroughly. Personnel should not re-enter the area unless given authority to do so by a responsible person.

4   A Safety Precautions card should be displayed close to the point of usage or storage of solvent.

Safe Working Procedures

Dipping

Dip cleaning is generally the most economical method of use.

1   Cleaning tanks should be at least three feet high or sited at bench height so that there is no danger of collapse into the tank.

2   Methods of cleaning which involve the operator leaning into the tank to remove articles by hand or to scoop out solvent must be avoided.

3   Tanks should be sited in well-ventilated areas away from direct draughts.

4   There should be sufficient freeboard (distance from liquid level to lip of tank) so that solvent vapour vapour remains in the tank and loss by evaporation is reduced. A general rule is to use a tank with freeboard depth equivalent to the shortest width of the tank.

5   Covers on dip tanks will help to reduce solvent in the atmosphere

6   Where possible work to be cleaned should be held in wire baskets so that the liquid can drain freely when the basket is removed from the solvent. Removal should be carried out slowly and items should be held in the freeboard until drainage is complete. Where heavy loads (10 kg or more) are being handled the use of mechanical handling equipment eg a simple hoist is strongly recommended.

7   Any movement of liquid over the surface being cleaned must be carried out within the confines of the tank.

8   Local or rim ventilation should be used for large tanks, say those with a surface area of exposed solvent greater than $1m^2$.

Wiping

Wiping is particularly suitable for cleaning large equipment which cannot be removed but the solvent should be used sparingly. Used cloths, containing solvent, should be stored in lidded containers.

Brushing

Brush cleaning is generally used to dislodge particles which cannot be removed by either dipping or wiping. Where possible the work to be cleaned should be placed in a tray to collect used solvent. Particular care should be taken to prevent solvent splashing into the eyes.

Spraying

Spraying is valuable for cleaning areas which are inaccessible to wiping and brushing - it may avoid the need to dismantle components for cleaning. However, spraying can lead to high solvent vapour concentrations, and close attention should therefore be paid to ensuring that ventilation is effective.

1   Wherever possible spray into a ventilated cabinet.

2   Use only an 'Airless' spraygun - injecting air into the solvent considerably increases solvent losses through vapourisation. Pressure-activated or electrically-activated spray guns should be used.

3 Use a solvent jet or coarse spray - unlike spray painting, large droplets are to be preferred. These dislodge the soil most effectively and reduce evaporation losses.

Equipment is shown schematically in Fig 1 which embodies these principles.

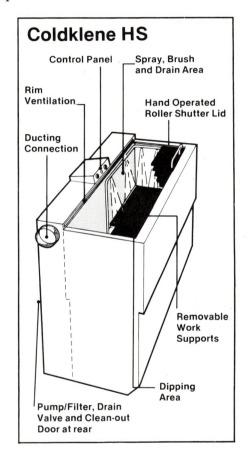

Fig. 1. Cold cleaning by dipping or jetting.

Atmosphere tests at operator nose level using this equipment to clean castings with 1,1,1-trichloroethane showed values between 50 and 120 ppm. The analysis was carried out using the Miran 104 portable gas analyser calibrated for the solvent.

3.2 Vapour Cleaning

The vapour cleaning process should only be operated in specially designed equipment whose key features include:

Appropriate materials of construction
Balanced heating and condensing capacity
Optimum freeboard to minimise losses by diffusion
Safety control devices to guard against condensing water failure, low solvent levels or overheating
Local ventilation at the rim of the plant

Significant new developments have taken place recently in the design of such open topped plants to include automatic lids, automatic solvent control and more effective rim ventilation (see Fig 2).

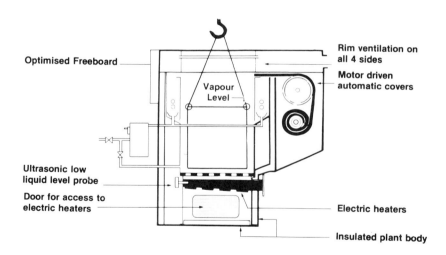

Fig. 2. S Series plant for vapour cleaning.

These and other features result in lower solvent losses and thereby lower worker exposure.

The increased use of mechanical handling is also important and ensure correct process control and minimum exposure.

Fig 3 shows the result of analyses carried out in the UK for a number of open topped vapour cleaning plants using trichloroethylene. These results demonstrate that with well designed equipment operated in accordance with the manufacturers instructions it is possible to work within the currently accepted TLV-TWA values. The analytical procedures used in this study have been described elsewhere and the results add to those already published.(Ref 1).

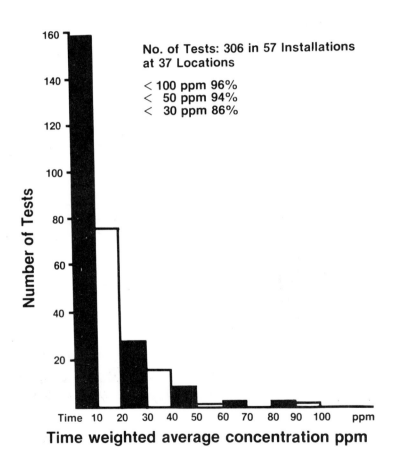

Fig. 3. Personal monitoring of trichloroethylene. Metal cleaning.

3.3 Drycleaning

Exposure measurements on workers operating drycleaning machines have also been carried out in the UK and Figs 4 and 5 show the results of these analyses for perchloroethylene and FC113 respectively. As for metal cleaning these results clearly show that minimum exposures can be achieved in this industry.

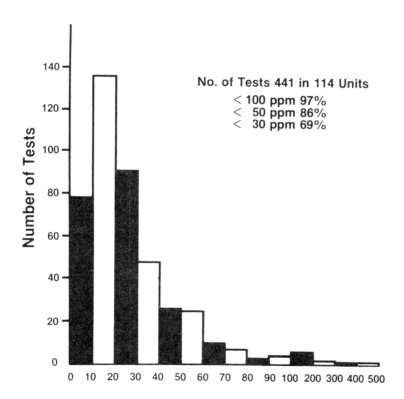

Fig. 4. *Personal monitoring of tetrachloroethylene. Drycleaning in unit shops.*

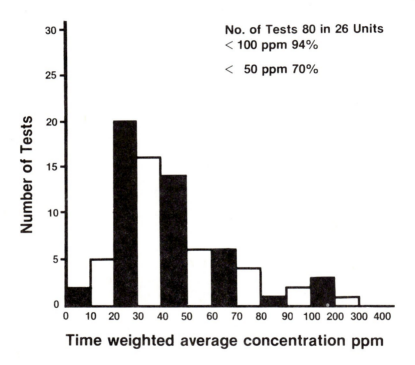

Fig. 5. Personal monitoring of 112-trichloro-122-trifluoroethane.

4 KEY PRINCIPLES AND CONCLUSIONS

4.1 Advice on Risks/Good Working Practices

Provision of advice on risks and good working practices to management and particularly shop floor workers is important. Since all volatile solvents present some risk it is important that the real risks are made known in literature, product data sheets etc. The consequences of failing to observe good operating practices must be explained to the actual user and precaution notices at the point of use can be valuable in this respect. Training packages are available for management to assist in this area but could be more widely used. One such package for vapour cleaning users also includes an aide memoire given to each trained man to help remind him of the key issues.

4.2 Process Design to Minimise Worker Exposure

Attention to operating methods in cold cleaning, to siting arrangements for cleaning processes and to the use of correctly designed modern equipment where possible using mechanised procedures are important. This approach ensures minimum exposure well within current guidelines. The success of this approach in the metal vapour cleaning and drycleaning industries shows what can be achieved.

4.3 Regular Maintenance Schedules

Cleaning methods should be reviewed regularly and not treated as an afterthought to production. Management attention in this area will help to avoid complacency and keep equipment and methods to a high standard.

4.4 Process/Worker Monitoring Procedures

A variety of techniques are available to estimate workroom concentration and worker exposure. Spot analysis using direct indicating tubes are valuable in checking the range in which the concentrations lie. Intelligent use of the data is important since either unnecessary alarm or complacency can result from analyses carried out at inappropriate sites in the production line.

The results shown in this paper for trichloroethylene tetrachloroethylene and trichlorotrifluoroethane have been determined by a variety of techniques. Adsorption of the solvent laden air onto carbon beds has been shown to be a reliable method (1) and has been compared with the carbon badge diffusive sampler method offered commercially by 3M Company and with the Miran IR continuous analysis technique. All give reliable results. For routine analysis the badge technique is clearly convenient and can be used to monitor worker exposure.

ACKNOWLEDGEMENTS

The author wishes to thank colleagues within ICI PLC, Mond Division, who were responsible for developing the appropriate techniques and carried out the experimental work and also to customers who co-operated in the personal monitoring study.

REFERENCES

1   Shipman A.J. and Whim B.P. (1980).   Ann. occup. Hyg. Vol 23 p.197-204.

ADVICE TO CUSTOMERS ON THE SAFE HANDLING OF SOLVENTS

R. A. Robinson

Occupational Health Group,
B. P. Chemicals Ltd., London

SUMMARY

Many solvents are of a hazardous nature and their distribution and safe handling are of prime concern to manufacturers and suppliers. It is therefore our policy, as well as our legal responsibility, to ensure that the customer understands the properties of the solvents we supply and is made fully aware of the handling procedures necessary for them to be used safely.

An internationally accepted regulatory framework is used to classify and label dangerous solvents and this provides the user, carrier and emergency services with an immediate assessment of the major hazards. Health and safety information is supplied on a standard format, known as the product Technigram, which gives details of handling precautions, emergency procedures, fire and explosion data together with a brief summary, where relevant, of the effects on the eyes, skin, following ingestion and on inhalation. New revelations on product health risks are also critically evaluated by experts and both customers and employees are advised if the findings are considered to constitute a new or increased hazard in normal use. In addition, sales and technical personnel receive a variety of customer queries on health and safety matters and these are processed by a formalised procedure, in co-operation with occupational health specialists, to ensure that the response given is effective and satisfactory.

These activities significantly exceed our statutory obligations but are now recognised as necessary to support the sale of solvents. Furthermore, they play an important part in enhancing our position in the market place.

1. INTRODUCTION

One of the essential tasks which accompanies the distribution of chemicals, many of which are hazardous, is to provide customers, carriers and emergency services with details of the properties of our products and give advice on their safe handling and use. This has been an integral part of BP Chemicals' marketing policy for many years and has become recognised as an important factor in promoting sales. Most countries now impose statutory obligations on manufacturers and importers which require them to make this information available when the material is placed on the market.

There are two main mechanisms within BP Chemicals for implementing these requirements for solvents :
a) ensuring that our solvents are clearly and adequately labelled
and
b) ensuring that the customer receives a written statement concerning the hazards of each solvent toegether with recommendations for its safety-in-use.

The latter information is included as part of the Company Technigram which is the prime document prepared for a specific product, or group of products, to support sales. In addition, we keep customers advised of new knowledge on the health risks of our solvents and sponsor studies to determine the relevance of such risks to the manner in which they may reasonably be expected to be used. An effective response to customer queries on health and safety matters is also necessary if genuine adverse effects of solvents we supply are to be correctly identified and controlled and unfounded allegations are to be dismissed.

2. LABELLING

Satisfactory labelling of solvents is particularly important as it provides an on-the-spot assessment of the major hazards of the product and indicates the chief precautions to be taken when handling. While the labelling of dangerous substances for transport is governed by a number of international agreements and conventions, legislation covering user labelling has only been promulgated on a similar level through Directives adopted by member states of the EEC. These regulations apply to a number of prescribed dangerous solvents when supplied in drums, but not when delivered in bulk directly into a storage tank or other vessel provided by the customer. However, recent resolutions passed by the United Nations and the Organisation for

Economic Co-operation and Development have indicated that these international bodies will soon become active in the general area of user labelling.

The EEC Directives have been implemented in UK law as 'The Packaging and Labelling of Dangerous Substances Regulations 1978 - Statutory Instrument No. 209' which set out the particulars which must be shown on drums, or any other form of packaging, which contain solvents or solvent mixtures listed as dangerous substances. Every container is required to show clearly and indelibly the following :
a) the chemical name of the solvent using an internationally recognised nomenclature,
b) the name and address of the manufacturer, the importer or the distributor of the solvent (the telephone number is usually given but this is not legally required),
c) the classification or danger symbol(s) designated to the solvent and an indication of the danger(s) involved in its use,
d) standard phrases, detailed in the regulations, indicating the special risks arising from such dangers,
   eg.   highly flammable
         harmful if swallowed
         may cause sensitisation
         irritating to the eyes and skin
e) standard phrases, detailed in the regulations, indicating the safety advice relating to the use of the solvent,
   eg.   keep out of reach of children
         do not breathe vapour
         do not empty into drains
         wear suitable protective clothing and eye/face protection.

In addition to listing the above information, the labels must be of a certain size (depending on the capacity of the container) with each danger symbol covering at least one tenth of the area of the label. The colour and presentation of the label shall also be such that the symbol and its background stand out clearly from it. BP Chemicals uses black print on a red and white background.

Although only having a legal obligation to label prescribed dangerous solvents, it is the Company's policy to label all its solvents using the same guidelines and criteria as described in the UK regulations. Danger symbols and risk phrases are omitted if they are not applicable, but appropriate safety phrases are given to provide handling instructions to both customer and user.

3. TECHNIGRAMS

These are essentially sales documents produced to provide existing and potential customers with important commercial and technical information on a specific product or group of products. It has long been the Company's practice to include some health and safety data in these technigrams. Over the last few years, however, legislation such as :

    UK   - Health and Safety at Work etc. Act (1974)
    USA   - Occupational Safety and Health Act (1970)
    Sweden - Work Environment Act (1978)

has been introduced which imposes for the first time a legal responsibility on manufacturers and suppliers of substances to provide such information to all its customers. These duties can be summarised as :

a) provide substances which are safe and without risk to health if properly used;
b) test and examine substances in order to be able to provide information on hazards and give advice on safe handling;
c) perform research to eliminate or minimise risks.

The detailed implications of these requirements have become clear since regulations were introduced in various countries and are now interpreted by BP Chemicals' advisers as being necessary to provide a package of health and safety information along the following lines :

<u>Storage and Handling</u> — Storage requirements; advice on unloading/transfer; precautions against specific problems; provision of adequate ventilation; relevant hygiene standards; personal protection procedures; instructions for disposal; relevant storage and transport regulations.

<u>Emergency Procedures</u> — Instructions on firefighting; first aid treatment, action on spillages.

<u>Fire & Explosion Data</u> — Statement on flammability/explosion hazards; auto-ignition and flammable limits; flash points; boiling points; whether miscible with water.

<u>Health Effects</u> — Odour threshold; effect of product on eyes, skin, by ingestion and when inhaled.

Current technigrams on solvents contain as much of this information as is considered appropriate. This will depend on the characteristic properties of the solvent and its envisaged applications.

4. UPDATING CUSTOMERS ON HEALTH RISKS OF SOLVENTS

Information on new health effects of existing products may be generated in a number of ways :

(a) **'In-House' Research**

This is usually carried out by the manufacturer to discharge his moral and legal obligations to carry out work to eliminate or minimise risks involved in normal use.

(b) **Informal Company Grouping**

A group of companies with a common interest in a particular chemical may participate in co-operative studies designed to define and evaluate a specific type of risk, eg. carcinogenicity.

(c) **Trade Associated Co-ordinated**

Assessment of commodity chemicals, particularly to determine the significance of long-term effects. have frequently been co-ordinated by trade associations such as the Chemical Manufacturers Association (in the USA) and the International Institute of Synthetic Rubber Producers. The European Chemical Industry Ecology and Toxicology Centre (ECETOC) have also become involved in this activity, particularly with the evaluation and interpretation of results. Costs have usually been shared based on nameplate capacity and use made of the trade association secretariat as the co-ordinating body.

(d) **Initiated by Regulatory Authorities**

In the USA and some other countries, studies are often initiated by the regulatory agencies who contract out specific toxicological investigations, usually based on advice from expert committees, to various laboratories.

(e) **Studies in Research Laboratories/Universities**

Many investigations, some rather poorly conducted, are carried out by independent research laboratories or in University departments by so-called experts. The programmes adopted are frequently influenced by the regulatory climate within a particular country and may well be politically motivated.

Immediately new information on the health effects of a particular solvent is published or becomes available, the manufacturer or importer of this substance has a duty to evaluate the results and advise their relevant customers if the findings indicate that a new risk is involved in normal use. In the USA certain provisions of the Toxic Substances Control Act make it an offence to withhold such information as may be considered to constitute a new health risk, and require the manufacturer to notify the results of the studies to the authorities without delay. This has resulted in many reports of health effects which are based on limited evidence and contain no indication of the relevance to man or to the industrial situation.

In Europe the regulatory scene is less restrictive and allows considerably more time to critically assess the significance of any new findings on the health effects of chemicals. In particular, since most studies are carried out on animals, every effort is made to determine the applicability of the results of such investigations to man, and to take into account the manner in which the substance is normally used, before communicating with customers.

When new information indicating possible health risks of a solvent marketed by a number of manufacturers is revealed, it is now recognised practice for experts from such manufacturers to meet together, agree on the interpretation of the results of the study, and prepare a common statement for their customers and their employees. This exercise is usually carried out in a co-operative spirit free from normal competitive attitudes. Furthermore, the UK has the additional facility of discussing any problem of this nature with the Employment Medical Advisory Service of the Health and Safety Executive both to obtain their views and consider the expediency of any joint action.

5. CUSTOMER QUERIES ON SOLVENT HEALTH RISKS

Sales representatives and technical service personnel will normally be the recipients of all customer enquiries about the health risks of our solvents. These can take the form of requests for general information on hazards/safe handling or concern about specific risks, for example, carcinogenicity or reproductive effects. Questions on the safety of current levels of workplace exposure and complaints of a medical nature may also be received. These have to be processed within the Company to ensure that a suitable and effective response is given and a formal procedure is in operation to achieve this.

5.1 General Information on Hazards/Advice on Safe Handling

Most simple queries for general information are handled directly by the sales function using the product Technigram. The first line of support for sales offices is the appropriate technical service department who have additional information and expertise relevant to their product range. Such departments may in turn seek assistance from occupational health advisers in the Company who can call for additional advice if necessary.

5.2 Information on Specific Hazards

Solvent producers are being increasingly asked to reply to specific requests on matters concerning the presence or absence of hazardous agents such as cadmium, specific carcinogens, or even carcinogens in general, in its products. Although these approaches may come from individual customers, they are frequently a reaction to legislative, trade union or media campaigns. Responses to such enquiries are normally co-ordinated by the appropriate technical service department in order to ensure that the information supplied is consistent and unambiguous.

5.3 Safe Levels of Workplace Exposure

Solvents which have well defined odours, or which cause irritation of the nose or eyes, are a frequent source of concern to customers because of the difficulty of differentiating between subjective perceptions and actual adverse effects. Caution is necessary before any reassurance can be given about the lack of such effects where there are subjective complaints. Wherever possible, the answer includes a recommendation that the customer takes steps to measure exposure levels and relate these to published hygiene standards. It may sometimes be necessary for technical service staff to evaluate exposures in customers' premises following this type of complaint. We also recognise the danger in providing an unsubstantiated reassurance that there is no adverse health effect, since this will fail to identify situations which do constitute a hazard.

Some customers may have their own resources for monitoring workplace exposure. Furthermore, we may give advice on relevant methods to be used and occasionally assist with the development of such methods. The significance of exposure will normally be established by relating measured levels to hygiene standards (usually Threshold Limit Values), which will normally be given in the technigram. It must be remembered, however. that published hygiene standards relate to personal exposures, normally over an eight hour period, and that measurements which do not adequately reproduce this are of doubtful validity and should not be relied upon as evidence of adequate control. The information supporting adopted hygiene standards is of variable quality and, while it is reasonable to assume that exposures in excess of a standard are a sound indication of the necessity for improved controls, firm reassurance based on lower levels can only be given where the biological effects of the solvent are adequately defined. All queries relating to atmospheric measurement are referred to the appropriate technical service group who seek assistance from the Company's occupational hygienist when confronted with an unfamiliar problem.

5.4 Clinical Problems

Complaints of adverse effects in employees using our solvents, or preparations containing our solvents, will invariably be a cause for considerable concern and it is essential that such complaints are quickly and adequately investigated. Occupational health experts are always advised at the earliest opportunity and, in order to facilitate an assessment of the problem, an attempt is made to obtain the following information from the customer at the time of the complaint:

a) The nature of the complaint including a detailed description of the reported clinical signs and symptoms. If the customer has his own medical or nursing staff, or where outside medical staff have been involved, their names, addresses and telephone numbers are obtained to allow direct medical consultation. Where practicable, the agreement of both the customer and the employees concerned are obtained to avoid subsequent difficulties on the disclosure of information.

b) The numbers of people affected and the total number of people exposed to the risk.

c) The time course of the complaint together with a summary indicating the start and finish dates of each individual complaint. Where large numbers are involved and the pattern of complaints is consistent, some form of numerical summary is requested.

d) Whether recent changes have been made in the processes, operating conditions and materials used. The relationships of such changes to the onset of complaints are particularly noted.

e) Whether steps have been taken to evaluate relevant exposure to our solvents or other materials used. Where

such information is available this should be supplied.
f) A general description of the processes used and the quality of engineering control of exposure as this will often provide useful information in relating these complaints to experience elsewhere.

Without the information detailed above, it is almost impossible to respond in a meaningful manner to a clinical problem and the exercise of collecting this data frequently enables the complaints to be seen in a new perspective. It is, for instance, sometimes found that an apparent complaint about a material used is the agreed compromise for some other industrial relations issue where there has been a desperate search for common ground between employer and his employees.

There will frequently be difficulties in obtaining information because of consideration of medical confidentiality and sometimes because of unwillingness on a customer's part to become involved in a thorough investigation of any problem. Direct contact between the doctors concerned can help in this situation and the offer of such consultation is offered where appropriate.

FACTORS AFFECTING THE USE OF FILTERING RESPIRATORS
FOR THE CONTROL OF SHORT-PERIOD EXPOSURES TO
HIGH CONCENTRATIONS OF ORGANIC VAPOURS

P. Leinster, M.J. Evans and M.F. Claydon

*The British Petroleum Co. p.l.c.,
Group Occupational Health Centre,
Chertsey Road, Sunbury-on-Thames, Middx TW16 7LN, UK*

1. INTRODUCTION

There are various short period job activities during routine operational and maintenance procedures when process containment is broken and for which it may not be reasonably practicable to control the release of organic vapour into the workplace atmosphere. If there is a potential for excessive exposure it is necessary to provide personnel with suitable respiratory protective equipment. In many such instances the tendency has been to specify air-line or self-contained breathing apparatus as this affords the greatest protection. Sometimes, however, the use of such protection is over-restrictive as well as being inconvenient and time-consuming. Therefore the use of filtering respirators to protect against short period, high concentrations of toxic organic vapours, such as benzene and ethylene oxide, during certain well-defined operations has been investigated.

There are basically three aspects to be considered in the specification and use of this type of respiratory protection:

1. The exposure situation should be defined both qualitatively (what contaminants are present?) and quantitatively (at what concentrations?) by appropriate occupational hygiene monitoring.
2. The performance capabilities of available respirators should be determined for the airborne contaminant(s) of interest either from manufacturers/suppliers information, or by testing, and the appropriate equipment selected.
3. Formal procedures should be written specifying when and how the equipment should be used

and adequate training given to those personnel involved.

Whenever there is doubt concerning either the qualitative or quantitative aspects of exposure, filtering respirators may be unsuitable and therefore breathing apparatus should be specified.

2. TYPES OF RESPIRATORY PROTECTION

In the United Kingdom the key documents regarding the specifications and use of respiratory protection are British Standards. BS4275: 1974(1) specifies canister respirators or self-contained breathing apparatus for toxic gas hazards and cartridge respirators for those gases of low toxicity.

Cartridge respirators, which are described in BS2091: 1969(2) are not recommended for use against materials having threshold limit values less than 100 parts per million, have a relatively low nominal protection factor of 20 and are therefore not suitable for the application under investigation. Canister respirators are described in BS2091 as being capable of removing limited concentrations of certain toxic gases, have a nominal protection factor of 400 and comprise a filter attached to a full facepiece. The filters have to pass an adsorption test which, for organic vapours, involves completely removing specified test gases (e.g. carbon tetrachloride) at 1% volume concentration for 30 minutes, at a flowrate of 16 ℓ/minute. The filters are quite large and are commonly attached to a harness or belt and connected to the facepiece via a tube.

However, BS2091: 1969 and BS4275: 1974 are showing their age for there is now a group of respirators, comprising a full facepiece into which the filter screws directly, which fall between the cartridge and canister types already described. These are known as "intermediate" or "screw-in" type respirators, have a nominal protection factor of 400 and are now marketed by most respiratory protective equipment suppliers. The filter capacity is generally less than that required to pass the British Standard canister adsorption test – although larger filters complying with this standard are now becoming available – and therefore the service life is probably shorter. However, many of these "screw-in" type respirators are designed without reference to any performance standard, although there is German Standard DIN 3181(3), published in 1980, that specifies performance characteristics for the filter; for example, the organic vapour filters should completely adsorb carbon tetrachloride at 0.5% volume concentration for 40 minutes at a flowrate of 30 ℓ/minute.

It is clear, therefore, that in principle both BS2091 type canister and the screw-in type respirators may provide a reasonable degree of protection against toxic organic vapours, with the screw-in type probably more convenient to use as the filter is directly attached to the mask.

3. PERFORMANCE REQUIREMENTS

For the short period job applications in mind, there was a requirement for a full facepiece filtering respirator to provide protection for 30 minutes against specified toxic organic vapours at concentrations up to approximately 1000 ppm. With a respirator there are two possible sources of leakage to be considered, namely through the filter and around the facepiece.

(a) *Filter Performance*

The service life of organic vapour respirator filters is determined by many factors including the nature of the contaminant(s), the concentration, work rate, humidity and temperature. Some manufacturers state in their literature that the wearer is warned in good time that the filter has reached capacity by smell/irritation (i.e. breakthrough of the vapour) experienced by the wearer, but with benzene, for example, the odour threshold of approximately 100 ppm is an order of magnitude higher than the Threshold Limit Value recommended by the American Conference of Governmental Industrial Hygienists. Therefore this approach is unacceptable and it was decided that an allowable service life should be defined for the contaminant of interest assuming standard conditions. The trapping of a contaminant by a filter material is a dynamic process and organic vapours are sorbed with different efficiencies determined by factors such as boiling point and polarity. The performance of a given filter material such as charcoal for a particular contaminant can only be determined from the adsorption isotherm found experimentally or estimated theoretically(4,5).

Carbon tetrachloride is specified for respirator filter testing in British, German and American standards. However, for example, if carbon tetrachloride has a breakthrough time of 100 minutes for a given sorbent the approximate breakthrough times for some other organic vapours are: benzene/hexane 80 minutes and dichloromethane 30 minutes. These figures indicate that using a standard material such as carbon tetrachloride for adsorption tests may be misleading as many materials break through more quickly. Thus wherever possible it is advisable to obtain breakthrough data for the contaminant of interest either experimentally or theoretically(4,5,6).

A range of performance tests have been carried out on various makes of filter and some of the results indicating the performance of one make of filter against ethylene oxide are given in Table 1. From a study of the data it is notable that the time to 1% breakthrough decreased significantly when the testing was carried out at the higher relative humidity. Furthermore, there was a 20% variation in the weight of sorbent in the filters. It was concluded

TABLE 1

Results of Respirator Testing

Ethylene Oxide, 850 ppm (v/v) Flow Rate 30 ℓ/min, Temperature 17-18°C			
Weight of Sorbent (g)	RH (%)	Time to Breakthrough	
		1% (Minutes)	10%
155	55	57	73
177	60	52	65
167	80	45	57
188	80	46	58

from these and other results that the allowable service life for this type of filter in circumstances where airborne concentrations up to 1000 ppm could occur should be 30 minutes, after which time the filter should be discarded.

(b) *Facepiece Leakage*

BS2091 specifies a face-seal leakage test in which the efficiency of a mask type is measured for ten clean shaven subjects, covering a broad spectrum of facial characteristics, using a sodium chloride aerosol. For a full facepiece to attain the specified nominal protection factor (NPF) of 400 the mean inward leakage for the ten subjects should not exceed 0.25%. Some actual face-seal leakage test data for a full facepiece are given in Table 2.

The results indicate that although the mask has a mean inward leakage of 0.25% (NPF=400) the individual results

TABLE 2

Face-Seal Leakage Test Data

Inward Leakage (%)
0.02
0.41
0.001
1.16
0.001
0.035
0.004
0.91
0.002
0.003
MEAN ($\bar{x}$) = 0.25
STANDARD DEVIATION (σ) = 0.41

vary by over three orders of magnitude. Moreover, by definition, 50% of wearers will receive less than the mean protection. The protection provided for ~85% and ~95% of wearers is given by the mean ($\bar{x}$) + the standard deviation (σ) and $\bar{x}$ + 2σ, corresponding to approximate nominal protection factors of 150 and 100 respectively. Thus to ensure protection for the majority of clean shaven subjects, it is recommended that a nominal protection factor of only 100 be assumed. Even so this allows use of the mask in concentrations up to 1000 ppm with a material having an occupational exposure limit of 10 ppm. However, the presence of beards or facial hair may reduce the level of protection achieved. These data set out in Table 2 indicate the importance of facepiece leakage and the necessity of ensuring that the best possible fit is achieved (most facepieces are available in different sizes) using at least qualitative fitting techniques, e.g. challenge with a pungent odour such as banana essence.

4. CONCLUSIONS

It was concluded that the screw-in type full-facepiece respirators could be used satisfactorily for certain well-defined short period exposure situations for which relatively high levels of organic vapour may be present provided that the following points are taken into account.

1. The potential exposure has been adequately evaluated.

2. Sufficient information is available either from the suppliers or from in-house testing to enable an allowable service life to be specified.

3. Facepiece fit is the limiting factor in the protection achieved, and the maximum nominal protection factor which should be assumed to protect the majority of clean shaven subjects is 100. It should be confirmed that this is adequate for each exposure situation.

4. Qualitative fitting techniques should be used to obtain as good a facepiece fit as possible. If a good fit cannot be achieved due to face shape, beards etc., the workers concerned should not carry out tasks involving the use of this type of respiratory protection.

5. Adequate training should be provided in the correct method of use and maintenance of the respirators.

5. REFERENCES

1. British Standard 4275: 1974. "Recommendations for the Selection, Use and Maintenance of Respiratory Protective Equipment".
2. British Standard 2091: 1969. "Specification for Respirators for Protection Against Harmful Dusts and Gases".
3. DIN3181. "Filters for Respiratory Protective Devices". May 1980, German Standards Institute.
4. Nelson, G.O. and Correia, A.N. (1976). *Am. Ind. Hyg. Assoc. J.* 37, 514-525.
5. Smoot, D.M. (1977). "Development of Improved Respirator Cartridge and Canister Test Methods". *DHEW, NIOSH Publication* No. 77-209.
6. Grubner, O. and Burgess, W.A. (1981). *Environ. Sci. Technol.* 15, 1346-1351.

THE USE OF SOLVENT SUBSTITUTION AS A METHOD FOR IMPROVING HEALTH AND SAFETY (1)

J. M. Ansell

Materials Safety Department
Administrative and Research Center
GAF Corporation
Wayne, NJ 07470

1. INTRODUCTION

Solvent substitution is one approach to reducing the risk associated with solvent usage. A wide variety of physical, chemical, and toxicological parameters should be considered is selecting a substitute.

The toxicological parameters most useful in determining solvent safety are discussed and a comparison of selected solvents made.

2. SOLVENT SAFETY

Organic solvents and their vapors are ubiquitous in most stages of chemical manufacturing. Because of this situation, organic solvents comprise the most important group of compounds which present an industrial health hazard. The United States production figures (Table 1) (2) of some of the more common halogenated hydrocarbons, show that solvents are high volume chemicals.

A substantial number of workers are exposed to solvents in their workplace, as well. The National Institute for Occupational Safety and Health estimates that as many as 1,900,000 workers were occupationally exposed to cyclohexanone in 1978 and as many as 500,000 workers were exposed to tetrachloroethylene in their workplace during 1979. When the entire U. S. population potentially exposed is considered, these numbers increase dramatically. Recent estimates (3) (Table 2) suggest 31,000,000 people in the U. S. are exposed to tetrachloroethylene at concentrations of greater than 0.4 $\mu g/m^3$.

TABLE ONE

U. S. PRODUCTION OF SELECTED HALOGENATED HYDROCARBONS IN 1979

Compounds	MM kg's
Carbon Tetrachloride	324
Chloroform	161
Methylene Chloride	288
Tetrachloroethylene	351
1,1,1-Trichloroethane	325

TABLE TWO

TOTAL POPULATION EXPOSURE FROM SPECIFIC POINT SOURCE EMISSIONS

Compounds	Population	Concentration Equal to or Greater Than $\mu g/m^3$
Carbon Tetrachloride	8,000,000	8.8×10^{-5}
Chloroform	826,000	1.0×10^{-2}
Methylene Chloride	1,680,000	1.0×10^{-2}
Phenol	20,500,000	1.0×10^{-4}
Tetrachloroethylene	31,000,000	4.0×10^{-1}
1,1,1-Trichloroethane	448,000	1.0×10^{-3}
Trichloroethylene	400,000	2.5×10^{-5}

When faced with a hazardous solvent in an occupational setting, there are three basic approaches one can use to reduce the risk. The first is to redesign the plant operations to isolate the area of solvent exposure. These engineering controls can either enclose the solvent, utilize ventilations, or a combination of both. This approach offers as advantages:

Effectiveness - exposure levels can be reduced to
   levels limited only by cost considerations.
Permanence - once designed and installed, constant
   supervision and restricted access will be minimized.
Process design - if the manufacturing is dependent on a
   specific solvent the continued use is allowed.

As disadvantages, engineering controls must be routinely monitored for effectiveness and maintenance; mechanical breakdowns are possible, and emission to the atmosphere may present a problem depending on local, state, or federal regulations.

A second approach is to utilize protective clothing and/or respiratory devices to isolate the worker. This approach is less expensive and simple to institute. It does, however, offer few of the advantages of the first method. Supervision is required to assure consistent and proper use of the equipment. Workers may find it cumbersome or uncomfortable. Atmospheric contaminants raise the concern of exposing employees in adjoining areas, or incidental exposure to employees in the area for a short time. The proper choice of equipment can be complicated and the useful lifetime difficult to monitor. Further exposure to chemical mixtures will require user validation of any clothing or device choosen.

The third approach is solvent substitution. Although not a panacea freeing one from all safety controls, it is extremely effective in reducing the hazards associated with specific operations.

Once the decision to utilize solvent substitution as the best approach to reducing risk has been made, one is faced with the problem of determining what is a "safe" solvent. This is often considered the risk to the health of workers. Certainly, this is an important concern but not the only factor to be considered. The usage hazards, including, but not limited to, vapor pressure, auto ignition temperature, flash point, flammability limits, and the hazard to a fire fighter in emergency situations, must also be reviewed.

There are, as well, many other factors that need to be considered in determining the overall safety profile of a solvent. These include bioaccumlation, biodegradability, effects on water purification systems, transport precautions, and storage requirements. Unfortunately, considering all of these are beyond the scope of this paper, which will focus on aspects relating to the health and safety of workers in an occupational setting.

It should be emphasized here that all chemicals, regardless of the intrinsic hazard, can be used safely. The concern, then, is how rigorous the handling requirements are to reduce the risk to an acceptable level. In fact, within this context, the rigorousness of handling requirements is a useful definition of solvent safety.

Of course, these requirements can only be defined in terms of specific operations. One process may require a high flash point. The potential for fire could be so great as to make the biohazard secondary. Similarly, an industrial process may be compatible with many available solvents so that reducing irritation may be sufficient to changing solvents. Others may find alcohol intolerance in workers, or an objectionable odor to workers (or adjoining neighborhoods), real "safety" problems.

3. TOXICOLOGY

The immense number of compounds presently used as solvents in large and small operations requires a narrowing of the number of materials reviewed in detail. For example, within the area of vapor degreasing and tank cleaning Table Three lists 12 compounds in 7 chemical classes which have found some application.

TABLE THREE

ALDEHYDES AND KETONES
   CYCLOHEXANONE
   FURFURAL
	CHLORINATED HYDROCARBONS
	   METHYLENE CHLORIDE
	   TETRACHLOROETHYLENE (PERCHLOR)
	   α-TRICHLOROETHANE
	   TRICHLOROETHYLENE (TCE)

AMIDES
   DIMETHYLACETAMIDE (DMAC)
   DIMETHYLFORMAMIDE (DMF)
   N-METHYLPYRROLIDONE (NMP)
	LACTONES
	   GAMMA-BUTROLACTONE (BLO)

AROMATICS
   PHENOL
	SULFOXIDES
	   DIMETHYLSULFOXIDE (DMSO)

Mutagens, Carcinogens and Teratogens

Based on the working definition used here for solvent safety, those compounds which have been cited as being mutagenic, carcinogenic or teratogenic would require quite stringent handling precautions. These are often beyond those necessitated by toxicology, but rather reflect the political climate surrounding these health effects today. It is not the purpose or intent of this paper to contribute to this discussion or to pass judgment on the testing which led to these citations. Suffice it to say, the handling of mutagens, carcinogens and teratogens imposes special handling requirements in today's environment.

Both tetrachloroethylene and trichloroethylene (TCE) have been cited as carcinogens. The EPA in its initial review concluded "evaluation of the data on these compounds presented substantial evidence that they were human carcinogens." (4)

If these evaluations are sustained after consideration of comments by EPA's Science Advisory Board, it is likely perchloroethylene (tetrachloroethylene) and trichloroethylene will be classified as high-probability carcinogens. (5)

In considering methylene chloride and 1,1,1-trichloroethane, the EPA's Carcinogen Assessment Group concluded "the weight of evidence led (them) to conclude in preliminary assessments that both chemicals exhibit suggestive evidence of human carcinogenicity." (5) It is important to note here that the EPA does recognize these designations are issues of considerable debate.

Phenol has been cited as a neoplastic agent and as a carcinogen by skin application in mice, although more recent work indicates it is a tumor promotor but not a carcinogen (6 a, b).

Dimethylacetamide has been cited as being teratogenic to mice by skin application as has dimethylsulfoxide to rats by oral administration, to rats and mice via intraperitoneal administration and hamsters by intravenous administration. (6 b)

The EPA in a review of dimethylformamide and n-methylpyrrolidone concluded the "literature raised concern over the chemicals teratogenic potential." (7, 8) In both cases, the increased embryonal mortality in DMF and the skeletal variations in NMP were observed only in the highest dosage levels and attributed to material toxicity. The EPA in its consideration of the data established a no-observed-effect-level of 554 ppm for DMF (7) and 1000 ppm for NMP. (8) These citations are summarized in Table Four.

TABLE FOUR

SUMMARY OF CITATIONS

POSSIBLE OR LIKELY	CONCERNED	NO PRESENT CONCERN
DMAC	DMF	CYCLOHEXANONE
DMSO	NMP	BLO
METHYLENE CHLORIDE		FURFURAL
PHENOL		
TETRACHLOROETHYLENE		
α-TRICHLOROETHANE		

Again, in the interest of brevity, a more complete comparison will be continued with DMF, NMP, BLO, furfural and cyclohexanone.

Acute Oral Toxicity

The first toxicological parameter which is usually considered in determining a compound's safety is the acute oral toxicity. However, I do not believe that LD_{50}'s are particularly good indicators of safety in an occupational setting.

First, acute ingestion of a large amount of solvent is not a significant route of exposure. Secondly, there can be large variations in the LD_{50}'s from one species to another. In the case of furfural an LD_{50} of 127 mg/kg is found when tested on rats but 2300 mg/kg is found for dogs. (9) Any real significance in this parameter is further complicated by toxicologists' inability to determine which species is most predictive of human response. There is, however, a regulatory reliance on these numbers, so they are listed in Table Five.

TABLE FIVE

ACUTE ORAL TOXICITY

Compound	Species	LD_{50} (mg/kg)
BLO	Rat	1500
Cyclohexanone	Rat	1620
DMF	Rat	2800
	Mouse	3750
Furfural	Rat	127
	Mouse	425
	Dog	2300
NMP	Rat	4200

Sub-Acute Feeding

Sub-acute feeding studies are more valuable in predicting the hazard to workers as they address longer term ingestion of smaller amounts of solvent. It is this type of exposure which is of concern in an occupational setting, rather than the single ingestion of a larger amount of solvent. Although lifetime studies would be the most useful parameter to judge, these studies are unavailable for a large number of compounds and often difficult to interpret in older studies. The three-month feeding studies then, are a compromise between availability and predictive ability.

Butyrolactone: Sub-acute oral feeding studies were carried out with BLO in rats and dogs. Both species ingested a nutritionally adequate diet containing BLO as 0.2, 0.4 and 0.8 percent of the diet over a 90-day period. Comparison to a control group fed the diet alone showed no adverse response in the parameters measured: behavior, appearance, survival, growth, food consumption, hemotology, clinical [chemical, urine analysis], or histopathologic evaluations. (10)

Cyclohexanone: Although specific 90-day feeding studies were not available cyclohexanone ingestion was predicted to lead to central nervous system depression. (11)

Dimethylformamide: The results of short-term oral studies conducted on mice, rats, rabbits and dogs have been summarized by the Food and Agricultural Organization and critiqued by R. Lauwery. (12)

A 90-day study conducted on male rats receiving DMF as 0.003 and 0.03 percent of their diets demonstrated no growth differences over controls. Mild hepatic steatosis was observed at the lower level and slight congestion of the kidneys was seen at the higher level.

Another 90-day study conducted on rats given DMF at .02, and .1 and .5 percent of the diet resulted in weight loss, reduced food intake and liver enlargement at the mid-dose level. At the higher concentration, anemia, leucocytosis, liver lesions, and hyperchloresterolaemia were seen.

Furfural: Furfural has a high degree of oral toxicity. Studies conducted on rats, dose and duration unknown, have produced cirrhosis of the liver. Little is known of the chronic effects although Central Nervous System effects have been reported. (9)

N-Methylpyrrolidone: A 90-day study conducted on Wistar-derived rats being fed NMP at concentrations of .08, 0.2 and 0.5 percent of their diet demonstrated no gross or behavioral abnormalities in any of the animals. There were no gross toxic or pharmacologic effects noted, nor differences in survival. In organ weight and clinical examinations, various minor, but statistically significant effects, were noted among the test group. These included increased male thyroid weights in the highest dose group, and differences in final urine pH and specific gravity in females. These effects were not toxicologically significant. There were no histopathological abnormalities observed attributable to the test diet. (10)

A 90-day study of mice fed NMP at concentrations of 0.04, 0.1 and 0.25 percent of the diet showed no difference in survival rate, or in gross or microscopic pathology. Clinical parameters were normal with the exception of elevated serum chloride levels. Spleen weight variations were noted in the highest dosage in females and in the mid and high dosage for males. (10)

Studies with dogs fed 0.1, 0.3 and 1.0 percent NMP in their diet for 90 days resulted in no change in mortality, body weight or food consumption, nor were gross signs of toxicity or behavioral abnormality noted. No gross or histological signs were found. The mid and high dosage levels resulted in some variation in clinical parameters but these were not found to have any associated toxicological relevance. (10)

Dermal Toxicity

A far more significant route of exposure to solvents is by skin. Therefore, dermal toxicity must be given careful consideration. Acute dermal toxicity data are shown in Table Six.

TABLE SIX

ACUTE DERMAL TOXICITY

Compound	Species	LD_{50} (mk/kg)
BLO	Guinea Pig	5600
Cyclohexanone	Rabbit	1000
DMF	Rabbit	5000
Furfural	Rabbit	620
NMP	Rabbit	>4000 <8000 ALD

Dermal Irritation

Dermal irritation studies are another toxicological parameter of concern in an occupational setting. Solvents are all expected to demonstrate irritation and dermatitis after prolonged and gross contact as they are excellent defatting agents. Physiological effects beyond this however should be given consideration. Repeated exposure to furfural has resulted in sensitization, allergic contact dermatitis, and photosensitization, which could present a significant usage hazard. (13) Additional DMF and Furfural are reported to be absorbed rapidly through the skin.

Ocular Irritation

These materials are all eye irritants and therefore this is not a significant parameter in distinguishing one solvent from another.

Inhalation Toxicity

The inhalation toxicity of a solvent is a strong indicator of its hazard as it is a significant route of exposure in an occupational exposure. In addition to animal studies, many reports are available on human experiences. Although this data is often subjective, with various degrees of supportive data it can be useful in setting a level of control.

Cyclohexanone: Cyclohexanone has TLV established of 50 ppm, although NIOSH recommends a 25 ppm level. Human volunteers report that exposure to 75 ppm for short duration results in eye, nose and throat irritation. Levels of 50 ppm were still found to be objectionable. These volunteers felt that 25 ppm was the highest satisfactory concentration for an eight-hour day.

The literature reports experiments where monkeys and rabbits were exposed to levels ranging from 190 ppm to greater than 3000 ppm for six-hour per day for 50 exposures. At the low levels slight kidney and liver damage was reported with increasing involvement at higher levels. (14)

DMF: DMF has an established TLV level of 10 ppm by the skin. People who have worked in an occupational setting where exposures range from 10 - 30 ppm have reported stomach pain, loss of appetite, nausea and headaches. Exposures of moderate concentrations for several weeks or a very high concentration for a short time, as may be experienced in the case of a leak, have reported similar symptoms, including throat irritation. Where workers were exposed to higher concentrations of DMF, they have reported alcohol intolerance with symptoms very similar to those of Antibuse compounds. Studies with dogs exposed at 50 ppm for six hours per day for three weeks, demonstrated reversible changes in blood pressure. Experiments at lower concentrations of 10 ppm for the six-hour day for an extended period of 107 exposures, resulted in more pronounced cardio-vascular effects; however, these were also reversible. The inhalation regimen of 23 ppm for 5-1/2 hours followed by 426 ppm for half hour for 58 days resulted in degenerative changes in the heart. This level resulted in the involvement of several other organs, including the liver and kidney. (12)

Furfural: Furfural has a TLV established of 5 ppm by the skin. Humans have reported eye irritation at levels as low as .15 ppm. People start perceiving the highly subjective measure of odor threshold at .25 to .4 ppm. In a chronic occupational setting with concentrations reported as higher than 1 ppm workers report headaches, irritation of the throat, lacramation, loss of taste, and tremors. (9)

The lowest lethal concentration for rats was measured as 153 ppm in a four-hour dosage. Cats, when exposed to 3000 ppm for one hour demonstrated salivation, paralysis, and cramps. This dose was fatal to one of 3 cats after three days. (13)

N-Methylpyrrolidone: No TLV has been set for NMP. The proposed TLV was intially between 100 and 200 ppm; however, when reviewed, the possibility of reaching this concentration was found extremely unlikely. Rats exposed to a saturated environment (greater than 500 ppm) for a single exposure of 6 hours had no effect when observed through two weeks. Similarly, no effect was seen at 370 ppm for six hours per day for ten days, or 120 ppm, seven-hour days, for 20 days. Studies on rats, guinea pigs, and rabbits at 50 ppm for eight hours per day for 20 days indicated no gross or histological abnormalities. (10)

4. CONCLUSION

In conclusion solvent safety must encompass all factors affecting the usage including transportation, usage hazards, toxicology, storage, and disposal. A safe solvent can be defined only in terms of a particular manufacturing operation and solvent substitution is an effective alternative which should be given careful consideration when faced with a hazardous solvent.

REFERENCES

1. Portions of this paper were first presented at the American Chemical Society, Division of Health and Safety, 181st National Meeting, Atlanta, Georgia, March 29 - April 3, 1981.

2. United States International Trade Commission Publication No. 1099, U. S. Government Printing Office, Washington, DC.

3. SRI International, (1979) "Human Exposure to Atmospheric Concentrations of Selected Chemicals" Office of Air Quality Planning and Standards, Environmental Protection Agency, Research Triangle Park.

4. 41FR 21402; May 25, 1976

5. 45FR 39768; June 11, 1980

6. a) Babich, H., and Davis, Dolo, (1981). Regulatory Toxicol. and Pharmacol., 1, 90-109.
   b) Lewis, R. J., and Tatken, R. L. ed., (1980) "Registry of Toxic Effects of Chemical Substances" U. S. Government Printing Office, Washington, DC.

7. 46FR15123; March 3, 1981.

8. 46FR151240, March 3, 1981.

9. N. Sax (ed.) (1980), Dangerous Properties of Industrial Materials Report, 1(2), 41.

10. GAF Corporation: Unpublished Data

11. Dreisbach, R. (1980) "Handbook of Poisoning: Prevention, Diagnosis, and Treatment" Lange Medical Publications, California.

12. R. Lauwery, (1976) Proceedings of the Meeting of the Scientific Committee, 12 December 1975, Carlo Erba Foundation, Occupational and Environmental Health Section, Italy.

13. Sittig, M., (1979), "Hazardous and Toxic Effects of Industrial Chemicals", Noyes Data Corporation, Park Ridge.

14. Rowe, U.K., and Wolf, M.A., (1963). **In** "Industrial Hygiene and Toxicology" (Ed. F. A. Patty). 2nd edn, Vol. 2, pg. 1765-1768. Interscience Publishers, New York.

EDUCATION AND TRAINING IN THE CONTROLLED USE OF SOLVENTS

Ralph G. Smith

School of Public Health, Department of Environmental and Industrial Health, The University of Michigan, Ann Arbor, Michigan 48109, USA

Industrial hygiene educational programs generally recognize solvents as a class of substances, and devote substantial amounts of time to considering the problems related to them. Certainly this is true of all graduate programs in industrial hygiene, as well as undergraduate curricula and short courses designed to familiarize non-professionals with industrial hygiene problems. To some extent this definition of the class of substances called solvents is unique in industrial hygiene curricula, for substances usually are classified according to their chemical or physical properties rather than their ultimate usage. Thus it is common to consider the toxic metals or the heavy metals as a class of substances having something in common, and similarly most curricula treat toxic gases as a class of substances of interest. All metals share certain chemical similarities, while all gases share the same physical state. Solvents, specifically non-aqueous solvents do share the same physical state, being liquids, but their consideration as a class of substances is related to their end usage rather than to this physical state or any chemical similarities. Solvents, in fact, consist of many diverse chemical substances with properties which may differ considerably.

In my opinion it is justified to treat solvents as a class of substances however and it is in fact quite useful to do so. In preparing this discussion I have elected to refer to the graduate curriculum in industrial hygiene at the University of Michigan with which I am most familiar. I would note however that this curriculum is very similar to that of other universities with similar programs in the United States, and I feel confident that curricula in other countries are indeed similar.

In Table I are listed all of the courses included in our curriculum which address problems related to solvents. The manner in which solvents are discussed in each course is perhaps self-evident, but a few words of explanation may be useful.

TABLE I

*University of Michigan Courses
In Which Solvents Are Studied*

- Fundamentals of Industrial Hygiene
- Air Sampling
- Industrial Hygiene Chemistry
- Industrial Hygiene Control
- Toxicology
- Safety

In the introductory course, Fundamentals of Industrial Hygiene, the general problems of usage, exposure, and control of solvents is logically addressed in one or two lectures. Next the air sampling and analysis course devotes a considerable percentage of available time to studies of the best methods for collecting samples of solvent vapors from the atmosphere and analyzing them in the laboratory. Alternatively direct reading instruments or devices for determining the concentration of solvents in the air are also considered. This particular subject is a most important one for prospective industrial hygienists, and the problems of sampling and analysis are therefore discussed in considerable detail.

The toxicology courses include among the many substances considered a number of solvents, and attempt to characterize classes of solvents as well as many individual members of these classes according to their acute and chronic toxicity characteristics. Clearly an understanding of the toxicity of individual solvents is absolutely essential.

Those courses dealing with the control of exposure to toxic substances, notably the industrial hygiene control methods course and the industrial ventilation course also include substantial reference to the problems of solvents in industry. Many examples of solvent usage are used including solvent vapor degreasing, spray painting and extractions. Other courses routinely included in graduate curricula include epidemiology, biostatistics, environmental law, etc. and exposures to solvents and solvent vapors are frequently used as examples of commonplace problems.

The content of the various courses should be of general interest, and the following discussion is therefore devoted to a consideration of the topics which ought to be addressed in a graduate industrial hygiene curriculum, and which for the most part are addressed in the University of Michigan curriculum.

Industrial hygiene is usually defined as a discipline which can conveniently and somewhat arbitrarily be divided into three categories of activity. Thus the official definition of industrial hygiene promulgated by the American Industrial Hygiene Association, states that:

"Industrial Hygiene is that science and art devoted to the recognition, evaluation and control of those environmental factors or stresses, arising in or from the work place, which may cause sickness, impaired health and well being, or significant discomfort and inefficiency among workers or among the citizens of the community".

The three key words in this definition are of course recognition, evaluation and control.

INDUSTRIAL HYGIENE - RECOGNITION

The recognition phase of industrial hygiene activity relative to solvents consists of developing an awareness of the properties of solvents as summarized in Table II.

TABLE II

Properties of Solvents

- Physical
- Chemical
- Toxicologic
- Hazardousness
    - Fire and Explosion
    - Health
        - Acute Effects, Inhalation
        - Chronic Effects, Inhalation
        - Skin Effects

In addition, it is essential to be thoroughly familiar with the processes in which solvents are used, the principal ones which are noted in Table III.

TABLE III

Processes In Which Solvents Are Used

- Coatings
- Extractions
- Degreasing
- Dry Cleaning
- Miscellaneous

Students are required to become familiar with these common industrial processes, both by information obtained from the scientific literature, and almost equally importantly by as much field experience and laboratory experience as possible. It is obviously of very great value to actually spend some time in a plant using a vapor degreaser, see the unit in operation and learn first hand of the health problems which can result from improper practices.

Next, students must become familiar with the physical, chemical and toxicologic properties of solvents. Here a systematic treatment of solvents classified according to their chemical nature is essential. It should be emphasized that economics plays an important part in the actual selection of solvents by industry, and that generally speaking the least expensive solvents, which are usually hydrocarbon mixtures, will be used whenever possible. Students must also be aware that the peculiar needs of the processes in which solvents are used further dictate the selection of solvents. Thus a particular mixture of solvents useful for the application of an enamel, for example may be quite different from a mixture of solvents used for the application of lacquers.

Knowledge of the physical properties of solvents is likewise important. By definition virtually all solvents are volatile, and are likely to possess boiling points within a specified range. Frequently the vapors are heavier than air, but the student must be cautioned against arriving at false conclusions concerning the density of the resultant vapor - air mixture with which he may actually be confronted. Such dilute mixtures, of course, usually have densities little different from that of pure air, and must be treated accordingly.

Considerable attention must be devoted in any curriculum to the toxicity of classes of solvents and certain individual members of each class. Consideration must be given to the effects of acute exposures, usually accidental, and chronic exposures extending over long periods of time. It should be stressed that although inhalation is the principal route of entry, certain effects may result from accidental ingestion, and many solvents are effectively absorbed by the intact skin. The tendency of all solvents to defat the skin and give rise to dermatitis problems should also be stressed.

Any industrial hygiene curriculum must, or certainly should include a course on safety engineering, and should stress the need for the industrial hygienist to be aware of the fire and explosion hazards related to the use of flammable solvents. Industrial hygienists as a group tend to neglect this important area and assume that others will take adequate precautions to prevent fires and explosions. Although such matters are indeed the responsibility of safety professionals, it is essential that the industrial hygienist be familiar with basic principles, if only to assist him in calling the attention of the safety professional to potential problem areas.

INDUSTRIAL HYGIENE - EVALUATION

Whenever solvents are used, and their potential hazards are recognized, the usual next activity performed by an industrial hygienist is evaluation of the extent of exposure. In Table IV are listed several kinds of activities, including the industrial hygiene measurements generally utilized in evaluating solvent vapor exposures.

TABLE IV

Evaluating Exposure to Solvents

- Analysis of Solvent Mixtures
- Air Sampling - Safety Related
- Air Sampling - Health Related
- Calibration of Sampling Equipment
- Biological Monitoring
- Medical Programs

Certainly the most commonly performed activity is that of air sampling to determine solvent vapor concentrations. This important activity should be afforded a substantial portion of courses concerned with air sampling and analysis.

It is probably correct to note that we are presently in a transitional period in which there is great enthusiasm for the use of simple and nearly foolproof methods of measurement as exemplified by the so-called "passive" monitors, or organic vapor badges. These devices require no skill on the part of the user but do require follow-up laboratory analyses. Students should be taught, however, that in many instances the use of such badges may not be entirely validated, hence conventional methods involving dynamic sampling will continue to be required. No doubt the most common of these is the use of small tubes containing activated charcoal or other adsorbing materials for collection and subsequent analysis. Other methods include the use of direct reading indicating tubes and direct reading instruments which are more or less specific for various solvent vapors. Similarly analytical procedures relying on thermal or solvent desorption of charcoal tubes followed by gas chromatographic analyses or other appropriate instrumental analyses should be stressed.

The great importance of frequently calibrating all equipment must be stressed, with reference not only to the calibration of pumps to confirm air flow readings, but also to the response of direct reading instruments to those materials which they allege to measure. Laboratory courses should include the preparation of known mixtures of solvent vapors and the measurement of the actual performance of several instruments. In addition, the bubble meter and devices similar to it should be utilized in experiments designed to teach methods of air flow calibration.

In many instances air sampling programs must be supplemented by biological monitoring of exposed workers for metabolites or other substances indicative of exposure, and in certain cases medical surveillance of employees may be required when specific effect can be observed which are believed to be related to solvent exposures.

SOLVENT VAPOR STANDARDS

On completion of air sampling or other evaluative procedures, students must be aware that the results should be carefully interpreted in the light of their relationship to existing governmental standards or other permissible exposure limits. Emphasis must be placed on the basis for such standards, noting that the standards are selected to prevent problems as simple as disagreeable odor levels, slight irritancy or possible early narcotic affects, and as serious as permanent and irreversible damage to target organs and possible cancer or mutagenic damage. It must be recognized of course, that if a legal standard is exceeded, then steps must be taken to ensure compliance regardless of the severity of the consequences of overexposure, but in many instances a thorough knowledge of the probable severity of existing problems will be of great value in determining the optimal course of action to be taken in implementing corrective procedures.

Standards designed to protect the environment should also be considered, for an awareness must be developed that the industrial hygienist cannot disregard the external environment when solving in-plant occupational exposure problems. Such matters are probably best treated in an elective or optional course concerned with air pollution and other environmental pollution problems, but if such a course is not available then at the very least students should be made aware of the relationahip between photochemical smog formation, and solvent vapor emissions. Likewise the environmental consequences of improper disposal of waste solvents onto land or waterways should be emphasized.

INDUSTRIAL HYGIENE - CONTROL METHODS

The final step in the overall process of evaluating exposures to solvents and their vapors, assuming that a problem has been found to exist is the application of measures designed to control exposures to acceptable levels. In this regard there is little that is unique to solvent vapors, and the conventional means of controlling exposures as summarized in Table V will generally prove to be successful.

TABLE V

Controlling Exposures To Solvents

- Packaging
- Labeling
- Storage - Transport
- Fire & Safety Measures
- Health-Related Measures
  - Substitution
  - Enclosure
  - Exhaust Ventilation

Achieving control by means of substitution is always an attractive means of proceeding, for it would appear to be a simple matter to recommend the use of "safe" solvent rather than one which is judged to be toxic or otherwise hazardous. It should be emphasized however that in practice substitutions are frequently not possible for a variety of technical or economic reasons, and furthermore that many solvents considered "safe" in the past have been demonstrated later to be less safe than believed. In general it should be stressed that control is likely to be achieved by the use of classical industrial hygiene engineering approaches, specifically the enclosure or containment of the process and the provision of well designed industrial exhaust ventilation systems. Good work practices used in conjunction with well designed control facilities are also essential and should be stressed. Prevention of direct contact of the hands or other parts of the body with solvents by means of suitably impermeable gloves, aprons, etc. should be discussed, with emphasis on current research demonstrating that many so-called impermeable substances are in fact of limited value in preventing the transmission of a solvent to the skin.

The general subject of the use of respiratory protection is of great importance and all curricula should treat this subject very thoroughly. The desirability of utilizing engineering controls rather than respiratory protection should be emphasized, together with the development of an awareness that there are many situations which will require the use of adequate respirators. Other specific topics to be covered include descriptions of the types of respirators available, methods of testing for good fit, and good respirator maintenance programs.

SPECIAL CONSIDERATIONS

All of the subjects discussed thus far apply to virtually every work situation where solvents may be used. There are several subject areas, however, which are somewhat more specialized and brief mention should be made of certain of these, as summarized in Table VI.

TABLE VI

Special Considerations

- Laboratory Usage Of Solvents
- Solvents Which Are Known Or Suspected
- Carcinogens
- Environmental Considerations

The use of solvents in ordinary laboratories, for example, is certainly a special case, for in most instances rather small amounts of a great many solvents may be used, and the rigorous application of industrial hygiene control procedures without modification may be impractical and unnecessary. For most laboratories, for example, it is quite adequate to control potential exposures to virtually any solvent by means of a program which first educates laboratory personnel, and requires that they rigorously adhere to work practice procedures requiring virtually all operations to be carried out in a properly operating laboratory hood supplemented by the wearing of adequate gloves to prevent skin contact. Pipetting by mouth should never be permitted, of course, and other more or less common sense practices should be respected. The term laboratory means many things, and in the case of pilot laboratories, relatively large quantities of solvents may be used in a manner quite analagous to actual manufacturing processes, hence such laboratories should be treated accordingly.

A number of solvents, some no longer used, and some still in use have been found to be carcinogenic, or at least are suspected of being carcinogens based upon high exposure levels administered to animals. Other solvents have demonstrated mutagenic activity, as evidenced by various tests. Depending upon the apparent potency of individual solvents in this regard, practical management decisions may argue for process changes in which the suspect solvents are no longer required, but frequently the data are not convincing, and the solvents continue to be used. In such cases it should be stressed that evaluation and control procedures are identical with those previously outlined, but common sense argues for extra efforts to minimize exposures to the greatest extent possible. In the special case of the very few materials which are classified as legal carcinogens, there are no options, and measures required by legal standards must be enforced.

Certain aspects of toxicity-related matters should be defined, including the special problems posed by mixtures, and the possibility that many substances dissolved in solvents may become more hazardous as a result of either synergistic action with the solvent or enhancement of the ability of the material to pass through the intact skin.

In summarizing this brief review of some aspects of an educational program dedicated to industrial health problems associated with solvents, it is apparent that the content of an adequate curriculum is very similar to the content of this Symposium. Certainly we have heard a number of interesting papers concerned with recognition, evaluation and control of solvent-related problems and the proceedings of this Symposium will be excellent material for both students and faculty to review in an effort to achieve an awareness of current problems and their possible solutions.

BIOLOGICAL TREATMENT OF ORGANIC COMPOUNDS AND SOLVENTS

D.A. Stafford

Department of Microbiology, University College, Cardiff, Newport Road, Cardiff, Wales, U.K.

1. INTRODUCTION

Organic compounds that find their way into natural water systems present a potential hazard and when these wastes are largely organic in nature, we need to think of the possible biological treatments that could be brought to bear on these waste compounds.

Many of them can be directly toxic to various forms of life in the ecosystem and any possibility of biological treatment is therefore of paramount importance. Very often, these compounds are present in waste waters from industry and they may be present in process waters. Since they contain a heavy polluting load, composed of an increasingly complex mixture of chemicals, their behaviour in biological systems can be very varied.

Starting in the summer of 1982 the second half of the 1974 Prevention of Pollution Act is progessively brought into effect and which will have profound influences within the U.K. on the treatment of such waste waters. It is of importance because of increasing environmental pressure to understand fundamentals when considering the biological treatment of organic compounds, particularly those compounds used as solvents in the chemical industry. In order to achieve maximum performance from the biological process it is necessary to examine the process at the microbial level. In any biooxidation system there are essentially two reactant components; firstly, the specific components within the waste itself and secondly the enzyme systems which are produced by micro-organisms, (Forster C.F., 1977). These micro-organisms are associated in biological treatment plants and they can either be found in the activated sludge unit system where the microbes are associated in flocs, or they can be in aerobic trickling filter systems where the microbes are attached to some physical support. The chemicals in a waste material are then trickled over this material support or pumped into an activated sludge unit and the microbes then oxidise the organic components. The microbes that may be involved in degrading these waste materials can be listed as follows:

Achromobacter
Acinetobacter
Comomonas
Flavobacterium
Pseudomonas

The way in which these micro-organisms can degrade organic solvent compounds and their analogues has been studied over the last twenty or thirty years fairly thoroughly, and an extensive understanding of the way which aromatic compounds and heterocyclic compounds are degraded is now available to us, (Callely, 1978; Cain, 1977).

2. BIO-DEGRADATION

It is important to understand the way in which we can measure the break-down of these organic compounds by microbial systems. Essentially, the compound is used by the micro-organism to promote its own growth. Often, however, a compound may be metabolised, but not completely, and sometimes without any obvious effect or benefit for the organism, (Hughes, 1977). Indeed, occasionally it may be to the detriment of the growth of the organism, (Jensa, 1963). Nevertheless, the microbe can in some way be involved in changing the chemical nature of the parent compound. When one considers the biodegradation steps involved, it is possible to gain the impression that this biodegradation is fairly straight forward. However, there are many different interpretations of the term 'biodegradation'. At one extreme it can mean the complete mineralisation of the compound by microbes to, say, carbon dioxide, sulphate, nitrate and water. On the other hand it can be used to describe a situation where the compound has only been altered, to some extent, slightly and that it has lost some of its characteristic properties. Another problem is that the intermediates that might be formed by the break down of the parent compound could indeed be more recalcitrant than the parent compound itself. So to environmentalists the definition of 'biodegradation' has not been very satisfactory. As a result of this standard methods have been put out by the Water Research Centre, (Stafford and Callely, 1977) and 'biodegradation' can now be identified in three different areas:

a. Primary biodegradation. This is taken to mean that the characteristic property(ies) of the original compound have disappeared and no longer respond to specific analytical tests. This definition does depend upon arbitrarily chosen criteria for defining the process and therefore lacks precise definition.

b. Environmentally acceptable biodegradation. This has been taken to mean the minimum alteration of the parent compound necessary to remove undesirable properties. This is a better definition but it similarly lacks precise definition.

c. Ultimate biodegradation. This is taken to mean the complete conversion of the parent compound to inorganic end products and products associated within the microbes normal metabolic processes. An aromatic compound might end up as, let us say, pyruvic acid or acetate within the cell and the compound then has been said to be ultimately biodegradable because these two possible biochemical intermediates are part of the cell's normal metabolism. This definition is intended to simulate the normal microbial breakdown in biological treatment systems. It must be remembered, however, that although the microbes degrade the compound to, let us say, carbon dioxide and water, a large part of the organic molecule does find its way through many aromatic compounds that are biodegradable and one example is given below, (Fig. 1) where we can see the breakdown of mandelic acid to acetyl Coenzyme A and succinate.

3. AROMATIC AND HETEROCYCLIC BIODEGRADATION

When some aromatic compounds are polymerised and found in polymers, they can similarly be degraded by microorganisms. One example is the degradation of coniferyl alcohol. The coniferyl alcohols can be produced in fact from lignin groups and they are converted to vanillic acid, and ether linkages between the various vanillic acid groups are broken, resulting in the liberation of vanillic acid itself, (Ander and Eriksson, 1978). Similarly other ether groups can be degraded to produce intermediates such as

Figure 1. The biodegradation of veratriglycerol-β (ortho-methoxyphenyl) by a <u>Pseudomonas</u> <u>sp</u>. coupled with a subsequent breakdown controlled by bacteria and fungi, (After Crawford <u>et al</u>., (1975) and Ander and Eriksson, (1978))

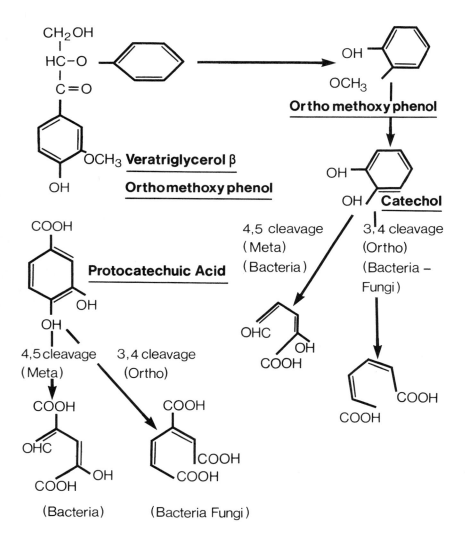

protocatechuic acid, (Ander and Eriksson, 1978). Once the
protocatechuic acids are formed they have to be further
degraded to intermediates, (Fig. 2).

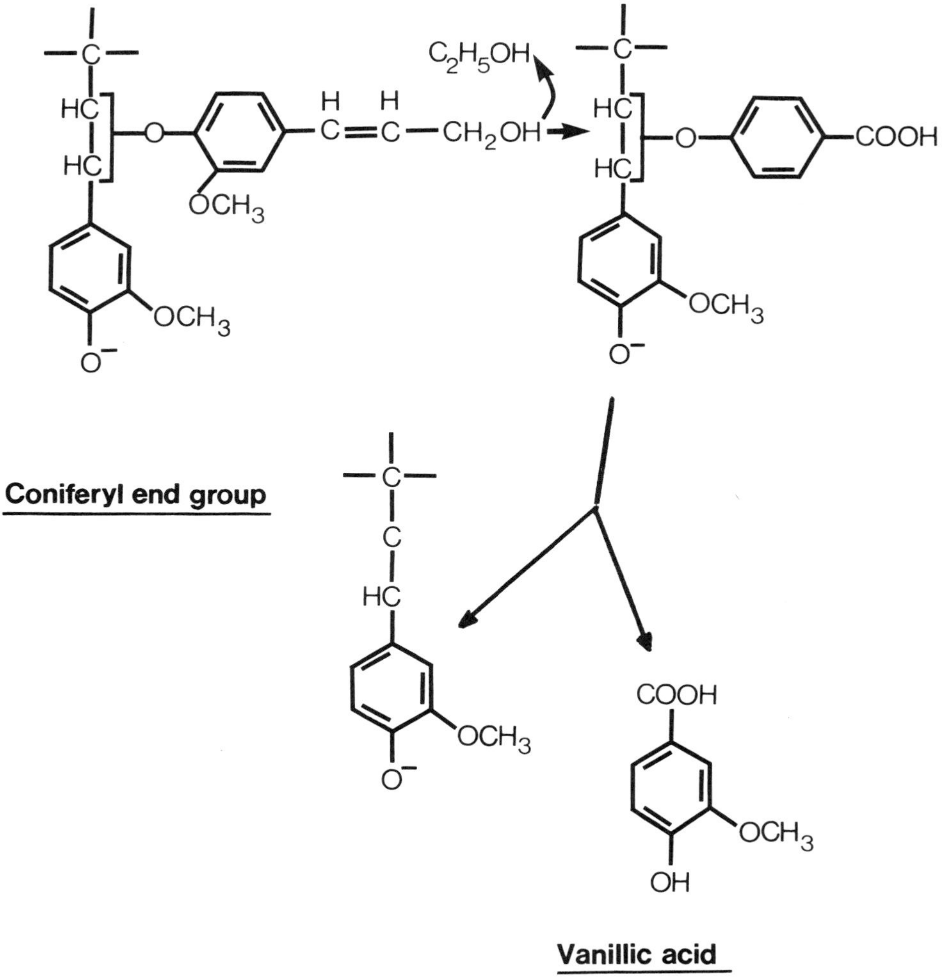

Coniferyl end group

Vanillic acid

Figure 2. Degradation of coniferyl end-groups (after
lignin residue breakdown) to vanillic acid by a (fungus)
Nocardia sp. (After Ander and Eriksson, 1978)

Other major groups that might be found in the production of drugs, synthetic herbicides, rubbers and so on and a number of them are shown in Figure 3.

QUININE   **NICOTINE**

a) **Drugs produced by living systems**

2- MERCAPTOBENZOTHIAZOLE

b) **Oxidant and vulcanization accelerator**

MORPHOLINE

c) **Anticorrosive**

Figure 3. Biologically degradable heterocyclic compounds

The range of compounds that can be biodegraded have remarkable chemical versatility. Many chemical solvents have such industrial applications that it is important to determine why, (if any), some are recalcitrant. If necessary it may lead to the need to use a biodegradable or non-toxic alternative.

The important thing with respect to the degradation of heterocyclic compounds is essentially the same as the degradation of aromatic compounds. It is important to metabolise the parent compound through intermediates such as tricarboxylic acid cycle. The primary degrading enzyme is known as an oxygenase which incorporates oxygen directly into the substrate molecule. If both atoms of the molecule are incorporated in this way, the enzyme is called a true oxygenase; if only one atom of the molecule of the oxygen is incorporated this is known as a mixed function oxygenase or hydroxylase. With hydroxylases there is also a requirement for a reducing agent to be present to convert the remaining oxygen atom to water. Of course ring sequences can be attacked in a number of sequences. They can be carbon-nitrogen, carbon-carbon, oxygen-carbon, carbon-sulphur, (Callely, 1978). Nicotine, for example, can be degraded via an open-ringed compound, (Fig. 4).

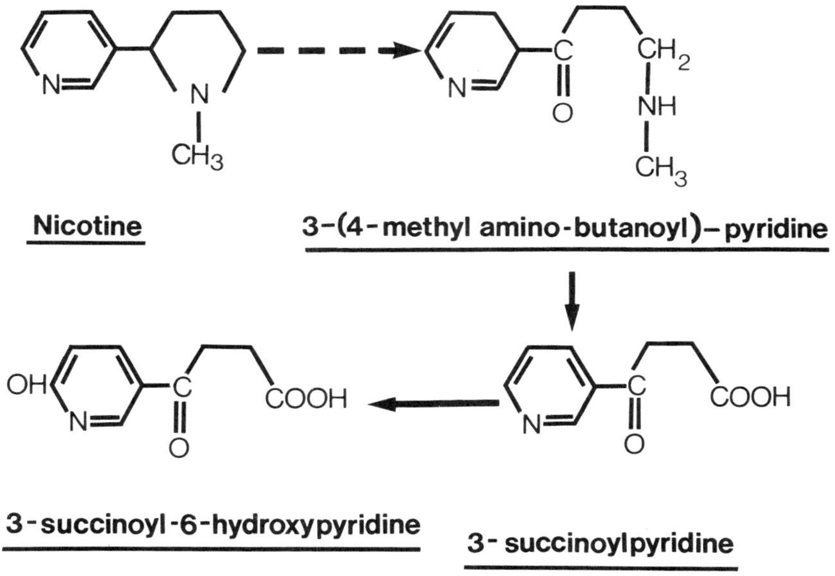

Nicotine   **3-(4-methyl amino-butanoyl)-pyridine**

3-succinoyl-6-hydroxypyridine   **3-succinoylpyridine**

Figure 4. Initial catabolic reactions for the microbial degradation of nicotine, (after Callely, 1978)

Indole can be degraded via anthranylic acid, (Fig. 5).

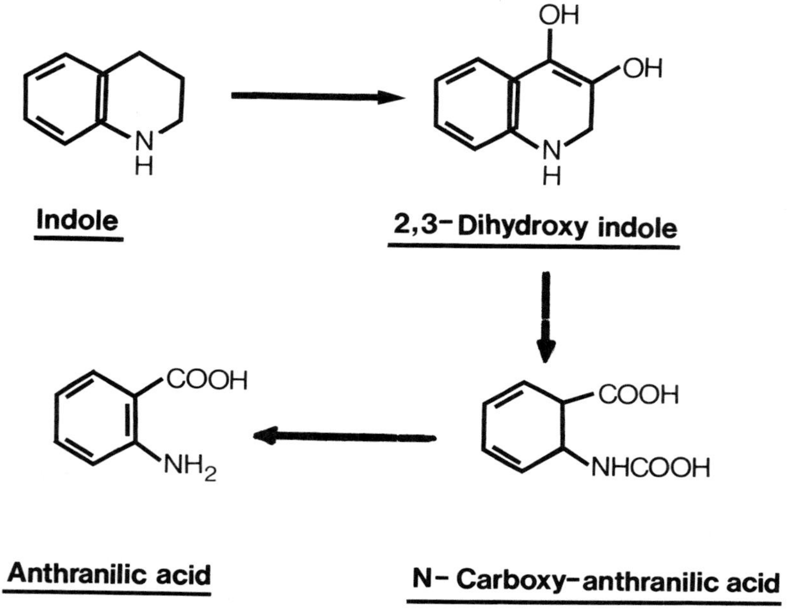

Indole → **2,3-Dihydroxy indole** → **N-Carboxy-anthranilic acid** → **Anthranilic acid**

Figure 5. Initial catabolic sequences for the microbial degradation of indole, (after Callely, 1978)

Other heterocyclics such as cyclohexanol and cyclopentanol are degraded to their lacto group and the ring can then be cleaved, (Fig. 6).

Figure 6. The microbial degradation of alicyclic compounds (eg. cyclohexanone and camphor), (after Callely, 1978)

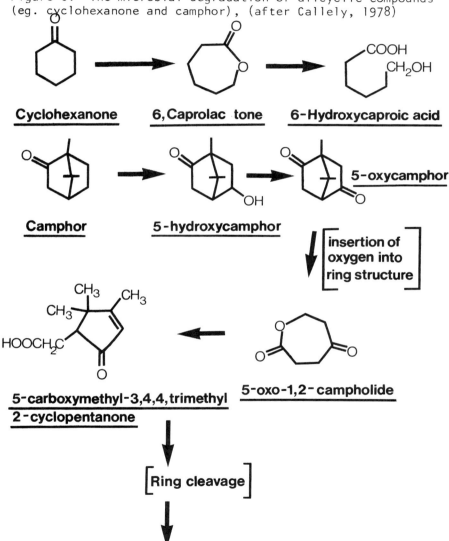

Sulphur compounds such as thiophenes found in oils and oil wastes can also be degraded by opening a ring structure, (Fig. 7).

Figure 7. Intermediate metabolic degradation of a Thiophene analogue (a component of crude oil fractions) using a Flavobacterium Sp. (After Callely, 1978)

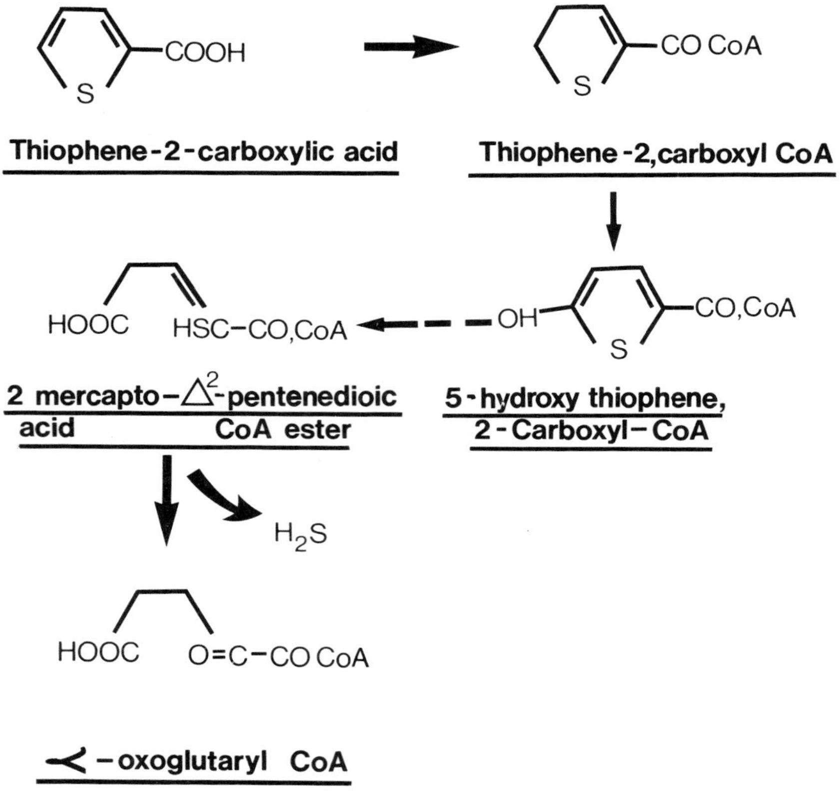

Thiophene-2-carboxylic acid → **Thiophene-2,carboxyl CoA**

↓

2 mercapto-Δ^2-pentenedioic acid   CoA ester ← --- **5-hydroxy thiophene, 2-Carboxyl-CoA**

↓ ↘ H_2S

HOOC   O=C-CO CoA

$\prec$-oxoglutaryl CoA

Many micro-organisms are available which can degrade a wide range of chemical compounds. For example, one microbe might be able to utilise phenol and many substituted analogues such as the methyl-phenols, the ethyl-phenols, the diethyl-phenols. In other cases one micro-organism may only be able to metabolise an individual compound. Certainly, in a mixed culture system one would expect that

the micro-organisms which are present and degrading a wide
range of organic compounds would be able to produce those
enzymes necessary for their biodegradation. Many of these
enzymes are specific for the particular compound, although
some enzymes are less specific, in that they are capable of
degrading not only an individual compound but some of the
analogues of it. One might consider in the future isolating
enzymes from microbes to degrade a particular compound or to
alter the structure of a particular compound such that its
undesirable characteristics have been removed. Such enzymes
might be fixed on a bed through which the compound can be
fed in solution. This immobilisation procedure can sometimes be difficult to maintain because of the breakdown of
the enzyme itself with time, but there are now many ways
for effectively immobilising intact organisms as well as
enzymes for industrial use, (Jack and Zajic, 1977).
Certainly this method might be much more efficient and
economic in terms of degrading particular compounds. This
method linked with the genetic engineering of microbes in
terms of cloning genes to produce particular enzymes, and
increasing their activity, might be one way of dramatically
improving the efficiency of treating these waste materials.
The simple way of improving the performance of a microbial
population is by the elective culture technique or by
acclimatisation of the micro-organisms over a period of
days, weeks, or even months, to biodegrade the compounds
present in the waste. It is important to give micro-
organisms sufficient time to adapt to these new wastes as
are being exposed to them so that the compounds can be
biodegraded in a particular environmental condition.
Sometimes, there are present other compounds during this
acclimatisation period which might lengthen the process
because they offer an alternative carbon source and energy
source for the micro-organisms. Such compounds could be,
for example, phenol in the presence of other substituted
aromatic and heterocyclic groups. In this case phenol would
be the preferred substance and the process in this
particular case would then be known as catabolite repression
where phenol represses those enzymes required to break
down the other compounds. One other problem too, is that
as soon as the compound is removed from the system then
the enzymes equired to break it down are then stopped being
produced by the micro-organisms. As such they lose
temporarily the ability to degrade that compound until the
compound is then re-introduced and more enzyme is produced.
In all of these microbial degradation processes it is simple
to remember the economy of the cell. Very often enzymes are

not produced that are not required to produce the energy for the cell, and where heterocyclic compounds are degraded the compound is used not only as an energy source but as a carbon source of both nitrogen and sulphur to build up amino acids and proteins. Whilst concentrating on the aerobic degradative pathways as we have done in this review, it is important to remember that many anaerobically grown microbes are also capable of degrading organic compounds and the phenolic groups have been largely studied in this particular way.

REFERENCES

1. Forster, C.F. (1977). Biooxidation. In Treatment of Industrial Effluents. Edtd. Callely, A.G., Forster, C.F. and Stafford, D.A. 65-87. Hodder and Stoughton, London.
2. Callely, A.G. (1978). The Microbial Degradation of Heterocyclic Compounds. In Progress in Industrial Microbiology, Vol. 14, 205-282. Edtd. M.J. Ball. Elsevier Scientific Publishing Co., Oxford.
3. Cain, R.B. (1977). Surfactant Biodegradation in Waste Waters. In Treatment of Industrial Effluents, 283-327. Edtd. Callely, A.G., Forster, C.F. and Stafford, D.A. Hodder and Stoughton, London.
4. Hughes, D.E. (1977). Microbes and Effluent Treatment. In Treatment of Industrial Effluents, 1-6. Edtd. Callely, A.G., Forster, C.F. and Stafford, D.A. Hodder and Stoughton, London.
5. Jensen, H.L. (1963). Acta Agre. Scand., 13, 404.
6. Stafford, D.A. and Callely, A.G. (1977). Microbiological and Biochemical Aspects. In Treatment of Industrial Effluents, 129-148. Edtd. Callely, A.G., Forster, C.F. and Stafford, D.A. Hodder and Stoughton, London.
7. Ander, P. and Eriksson, K.E. (1978). Lignin degradation and Utilization by microorganisms. In Progress in Industrial Microbiology, Vol. 14, 1-58. Edtd. M.J. Ball, Elsevier Scientific Publication Co., Oxford.
8. Crawford, R.L. (1975). Can. J. Microbiol. 21, 1654.
9. Jack, T.R. and Zajic, J.E. (1977). Adv. Biochem. Eng. 5, 125.

ENVIRONMENTAL MONITORING OF SOLVENT EXPOSURE

M. Fugaš

*Institute for Medical Research
and Occupational Health,
Zagreb, Yugoslavia*

1. INTRODUCTION

Industrial solvents are by definition "fluids or mixtures of fluids that are capable of dissolving or dispersing other substances to produce compositions of industrial value" (1). Now, however, many products formerly used as solvents find their main application as chemical intermediates, as components of antifreeze compositions or hydraulic fluids or in other non-solvent utilizations. On the other hand as to their properties and behaviour in the environment they belong to the family of volatile organic compounds and what applies in general to the whole group is also true for the solvents. Therefore the environmental monitoring and assessment of solvent exposure will be discussed within a broader frame of volatile organic compounds produced by human activities. Within this context under "exposure" a non-occupational exposure of general population will be referred to.

According to the available data estimates of hydrocarbon emissions from human activities expressed in million tons per year are shown in Table 1. Estimated emissions from mobile sources amount from 26 to 70 per cent of the sum and out of these 99 per cent come from petrol-powered vehicles. Since mobile sources are most widespread - their emissions are released to the pedestrian level in the densely populated areas - and since they take part in photochemical reactions producing even more objectionable compounds affecting human health, vegetation and local climate, in the recent past

TABLE 1

Estimates of hydrocarbon emissions from human activities expressed in million tons per year

Area or country	Reference	Year	Sources mobile	stationary
World wide	2	1968	34	54
USA	2	1969	19.8	17.6
West Germany	3	1975	0.76	1.05
Ruhr district	4	(1980)	0.0137	0.0387
Sweden	5	1975-78	0.181	0.250
UK	3	1975	0.660	0.665
Kanagawa Prefecture*	6	1973-74	0.083	0.036
Harima area*	7	1980	0.0105	0.0105

*Japan

closest attention was paid to the environmental pollution by hydrocarbons coming from motor vehicles.

The most extensive measurements of primary and secondary air pollutants coming from car exhausts were made in southern California (8), where type of pollution first reached the critical level.

In the past the presence of volatile organic chemicals in the environment coming from sources other than motor vehicles was generally considered more as a nuisance perceptible by odour and/or taste than as a serious health hazards to general population. After a number of these compounds were suspected of mutagenic, teratogenic and carcinogenic effects an interest has arisen in environmental pollution by volatile organic compounds.

A workshop panel assembled by the U.S. National Science Foundation in Washington has selected 80 compounds, out of a list of 340, considered to be hazardous to the environment and human health, as a priority list for further research (9). The VDI guideline for emission control of organic compounds especially solvents (10) includes a list of 138 chemicals which can be expected in the air.

Thanks to the advances in analytical and sampling techniques in the last twenty years identification and measurement of very low concentration levels of volatile organic compounds was made possible so that inventories of organic pollutants in environmental media could be made, the pollutants related to the sources, and their behaviour and fate studied (11-14). In this way a catalogue of organic substances to which humans might be exposed has been obtained.

Exhaustive reviews on the measurement of organic compounds in air and environment have been published recently (15,16,17). Detailed information on the methods for the determination of volatile organic compounds in water can be found in the proceedings of the first and second European symposium "Analysis of Organic Micropollutants in Water" (18,19), and therefore only those aspects which are specific for exposure monitoring will be discussed.

2. MONITORING FOR EXPOSURE ASSESSMENT

The first step in environmental monitoring is to develop a sampling strategy which will give representative and meaningful samples. For human exposure assessment samples should be taken from the air people breath, from the food people eat and from the water people drink. In countries with organized water supplies to obtain representative samples of drinking water should not be a problem. It is less simple to obtain representative samples of food - the samples of daily meals ready for consumption and collected in households of investigated population. To collect representative samples of air is even more complicated and therefore I shall pay more attention to it.

It has been shown that in the average people living in towns spend 5 to 10 per cent of their time outdoors where most measurements take place. The rest they spend indoors: at home 50 to 60 per cent, at work 25 per cent, in public buildings, shops, restaurants, etc. 5 to 15 per cent of the time (20). Thus, in addition to being exposed to solvent emissions from industrial plants, petrol stations, evaporation losses and exhausts from motor cars, fuel combustion and dry-cleaning shops, people are also exposed to evaporations of an ever growing variety of products for everyday use such as building materials and surface coatings for walls and furniture, cleansing agents, shoe polish, insecticides, deodorants, hair sprays, etc.

There are two ways of assessing people's actual exposure (21,22): 1) by measuring personal exposure over a representative period of time or 2) by calculating time weighted average exposure from concentrations measured in the air of each microenvironment in which people are likely to stay, and from the records on the amount of time they actually spent in each microenvironment. One or the other way has been used to assess personal exposure of people to a number of air pollutants (23, 24).

For measurement of personal exposure of general population, personal samplers, first designed for measuring occupational exposure (25), have been adopted (26). The battery operated samplers cannot work continuously over 24 hours. Therefore these samplers mostly used while people are moving around, are substituted by stationary samplers at home or during sedentary work. Pumps have to be calibrated and charged and only a limited group of people willing and capable to carry and maintain the samplers can be monitored. Passive samplers have opened a possibility of much wider application of personal monitoring (27).

The time weighted average model based on "microenvironment" approach is less dependent on the cooperation with individuals - other than operators. Microenvironments can be selected on the basis of a simple questionnaire or a more through time-motion study (28). It is a three-dimensional model observing people moving through time occupying different microenvironments and exposing themselves to the particular concentration associated with each microenvironment in the given period (26, 21,28).

A further step forward is a computer-simulation model of human exposure to air pollution following the exposure of each single person in time and space (29). The integrated exposure E_i of a person $\underline{i}$ is computed as the sum of the products of the concentrations c_j encountered in each microenvironment $\underline{j}$ and the time t_{ij} spent there:

$$E_i = \sum_{j=1}^{J} c_j t_{ij}$$

where J is the total number of microenvironments occupied by person $\underline{i}$ in some period of interest.

3. BIOLOGICAL INDICES OF SOLVENT EXPOSURE

Although personal exposure monitoring can provide much more truthful data on individual exposure to air pollution than fixed monitoring stations, the risk from exposure to organic solvents is more directly related to solvent uptake (30). The uptake depends on respiratory fitness, respiration rate, the amount of stored adipose tissue, genetic and environmental factors. It may be estimated by measuring the solvent or its metabolites in the breath, blood or urine depending on the metabolic fate of the solvent under consideration.

Verberk and Scheffers (31) have measured tetrachloroethylene in alveolar air of 136 people living near 12 dry-cleaning shops and found that the mean concentrations decrease with the distance from the shops from 4.9 to 0.22 mg/m^3.

4. HUMAN EXPOSURE VERSUS BODY BURDEN

Wallace and co-workers (32) carried out a pilot study on human exposure and body burden for 15 volatile organic compounds. The study objective was to field-test the methods of measurement and to

TABLE 2

Estimated daily intake* of 10 volatile organic compounds through air and water for 17 subjects (µg/day) (32)

Subject	Air	Water	Total	Percent from Air
30001	12,400	200	12,600	98
30002	2,700	150	2,850	95
30003	12,000	600	12,600	95
30004	400	150	550	73
30005	300	150	450	67
30011	550	140	690	80
30012	1,250	160	1,410	89
30013	9,000	160	9,160	98
30014	300	150	450	67
30015	2,600	140	2,740	95
30016	1,140	130	1,270	90
40001	2,470	280	2,750	90
40002	2,390	260	2,650	90
40003	1,250	240	1,490	84
40011	300	240	540	56
40012	1,240	230	1,470	84
40013	3,800	210	4,010	95

*Assuming 10 m^3/day and 1 liter/day intake rates for air and water

compare the volatile organic compound levels in the breathing zone air and drinking water with the levels of the same compounds in human breath.

The study was conducted in two areas: a petrochemical manufacturing centre in Texas and a non-industrial community in North Carolina. Subjects, 11 from Texas and 6 from North Carolina, were volunteers from local universities with no direct contact with organic chemicals either at university or in hobbies or occupations. The participants carried around air monitors, consisting of a Tenax cartridge and a small pump, for a 5 to 9-hour period while carrying out their normal daily activities.

TABLE 3

Spearman correlation coefficients between air and breath for estimated levels of selected vapor phase organics - both groups (32)

	n		$r_{Air-Breath}$	
	UNC	Lamar	UNC	Lamar
Benzene	5	11	.70	.04
Chloroform	5	11	.60	.20*
Vinylidene chloride	6	11	.48	.67
1,1-Dichloroethane	6	11	.00	.00
1,2-Dichloroethane	6	11	.44	.07*
1,1,1-Trichloroethane	6	11	.94**	.63
Trichloroethylene	6	11	.94	.41
1,2-Dichloropropane	6	11	.00	.00**
Tetrachloroethylene	6	11	.20	.80
Bromodichloromethane	6	11	.20	.00
Dibromochloromethane	6	11	.00	.00
Ethylene dibromide	6	11	.00	.00
Chlorobenzene	6	11	.31	.00
Dichlorobenzene isomer	6	11	.03	.08
o-Dichlorobenzene	6	11	.00	.00

* $p<.01$ ** $p<.05$

They filled a water sample vial each time they had a drink. At the end of day air monitors and water samples were returned and breath samples were collected on Tenax GC cartridges by means of a specially designed spirometer.

All samples were analysed by capillary column gas chromatography-mass spectrometry. Air and breath samples were transferred by thermal desorption and water samples by a purge and trap technique (33). Electron capture and flame ionization detectors were operated simultaneously to detect both benzene and halogenated compounds.

An extensive quality control programme was carried out simultaneously.

The results have shown that simultaneous direct measurements of individual human exposure to a greater number of organic compounds are feasible. A great variability of exposures in a homogeneous group of subjects was observed (Table 2) and an apparent relationship between inhaled and exhaled concentrations of several compounds (Table 3). On the basis of this first exploratory study additional studies have been undertaken to extend the described monitoring approach to a statistically valid sample of several hundred people in an industrial community.

5. CONCLUSION

Very little is known about the exposure of general population to solvents. The most recent data seem to show that methods of measuring human exposure to and body burden of volatile organic compounds are now available and studies in progress are expected to give information on qualitative and quantitative aspects of this problem.

Even less is known about the contamination of the environment by solvents. Only accidental spills of petroleum hydrocarbons and their effects (which were evident) have been well described and quantified. They, however, make up less than 10 per cent of the total amount of oil that man introduces in the world's waters (34). The study of pollution levels which are due to normal everyday practices and of possible long-term effects of solvents on the human environment are highly desirable.

REFERENCES

1. Solvents, Industrial (1954). *In* "Encyclopedia of Chemical Technology" (Ed. K. Othmer), Vol. 12, pp. 654-685. Interscience Encyclopedia Inc., New York.
2. National Research Council, Division of Medical Sciences, Committee on Biologic Effects of Atmospheric Pollutants (1976). "Vapour Phase Organic Pollutants". National Academy of Sciences, Washington D.C.

3. Bruckmann, P. (1981). *In* "Proceedings of the Second European Symposium Physico-chemical Behaviour of Atmospheric Pollutants" (Eds. B. Versino and H. Ott), pp. 336-348. Commission of European Communittees. D. Reidel Publishing Co., Dordrecht.
4. Ministerium fur Arbeit, Gesundheit und Soziales des Landes Nordrhein-Westfahlen (1980). "Luft-Reinhalteplan Ruhrgebiet Mitte 1980--1984". Düsseldorf.
5. Aswald, K. (1981). The National Swedish Environmental Protection Board, Personal Communication.
6. Industrial Pollution Control Association of Japan (1975). "Study on Emission Control of Hydrocarbons". Tokyo.
7. Inagaki, K. (1981). Industrial Pollution Control Association of Japan, Personal Communication.
8. U.S. Department of Health, Education and Welfare, Public Health Service, National Air Pollution Control Administration (1970). "Air Quality Criteria for Hydrocarbons", NAPCA Publication No AP 64. U.S. Government Printing Office, Washington D.C.
9. Stephenson, M.E. (1977). *J. Ecotoxicol. Environ. Saf.* $\underline{1}$, 37-48.
10. VDI-Kommission Reinhaltung der Luft (1977). "Organische Verbindungen - insbesondere Losungsmittel". VDI Richtlinie 2280. Düsseldorf.
11. Williams, I.H. (1965). *Anal. Chem.* $\underline{37}$, 1723--1732.
12. Okita, T., Watanabe, M. and Kiyono, S. (1970). *J. Jap. Soc. Air Pollut.* $\underline{5}$, 73 (in Japanese).
13. Grob, K. and Grob, G. (1971). *J. Chromatogr.* $\underline{62}$, 1-13.
14. The European Community Data Bank for Environmental Chemicals.
15. Leinster, P., Perry, R. and Young, R.J. (1977). *Talanta* $\underline{24}$, 205-213.
16. Budde, W.L. and Eichelberger, J.W. (1979). *Anal. Chem.* $\underline{51}$, 567 A-574 A.

17. Lamb, S.I., Petrowski, C., Kaplan, I.R. and Simoneit, B.R.T. (1980). *J. Air Pollut. Control Assoc.* **30**, 1098-1115.
18. Commission of European Communities (1979). "European Symposium Analysis of Organic Micropollutants in Water". Berlin.
19. Commission of European Communities (1981). "Proceedings of the Second European Symposium Analysis of Organic Micropollutants in Water". (Eds. A. Bjorseth and G. Angeletti). D. Reidel Publishing Co., Dordrecht.
20. Fugaš, M., Šega, K. and Šišović, A. (1982). *Environ. Monit. and Assess.* **17**, in print.
21. Duan, N. (1981). "Models for Human Exposure to Air Pollution". Paper presented at the International Symposium "Indoor Air Pollution, Health and Energy Conservation". Extended Summaries. Harvard University, Amherst, MA.
22. Fugaš, M. (1975). Paper 38.5. *In* Proceedings of the International Conference on "Environmental Sensing and Assessment". Vol. 2, Institute of Electrical and Electronic Engineers, New York, NY.
23. Harvard University (1981). International Symposium "Indoor Air Pollution, Health and Energy Conservation". Extended Summaries. Session E-2 "Exposure Studies", Amherst, MA.
24. National Research Council, Board on Toxicology and Environmental Hazards, Committee on Indoor Pollutants (1981). "Indoor Pollutants", pp. 269-301. National Academy Press, Washington D.C.
25. Sherwood, R.J. and Greenhalgh, D.M.S. (1960). *Ann. Occup. Hyg.* **2**, 127-132.
26. Fugaš, M., Wilder, B., Pauković, R., Hršak, J. and Steiner-Škreb, D. (1973). *In* the Proceedings of the International Symposium on "Environmental Health Aspects of Lead", pp. 961-968. Commission of the European Communities, Luxemburg.

27. West, P.W. and Reiszner, K.D. (1979). *In* the "Proceedings of the Symposium on the Development and Usage" of Personal Monitors" (Eds. D.T. Mage and L. Wallace) pp. 461-471. U.S. Environmental Protection Agency. Report No EPA 600/9-79-032. U.S. Government Printing Office, Washington D.C.
28. Moschandreas, D.J. and Morse, S.S. (1979). "Exposure Estimation and Mobility Patterns". Paper No 79-14.4 presented at the 72nd Annual Meeting of the Air Pollution Control Association, Cincinnati, Ohio.
29. Ott, W.R. (1980). "Concepts of Human Exposure to Environmental Pollution". SIMS. Technical Report No 32. Stanford University, Department of Statistics, Stanford, Cal.
30. Gompertz, D. (1980). *Ann occup. Hyg.* $\underline{23}$, 405-410.
31. Verberk, M.M. and Scheffers, T.M.L. (1980). *Environmental Research* 21, 432-437.
32. Wallace, L. and Pellizzari, E. (1981). "Individual Human Exposure to Volatile Organic Compounds Encountered During Normal Daily Activities, paper presented at the International Symposium "Indoor Air Pollution, Health and Energy Conservation". Extended Summaries. Harvard University, Amherst, MA.
33. Bellar, T.A. and Lichtenberg, J. (1974). *J. American Water Works Assoc.* $\underline{66}$, 739-744.
34. Massachusetts Institute of Technology (1970). "Mans Impact on the Global Environment". Report of the Study of Critical Environmental Problems. The MIT Press, Cambridge, MA.